AF344272

HAFNIUM

CHEMICAL CHARACTERISTICS, PRODUCTION AND APPLICATIONS

CHEMISTRY RESEARCH AND APPLICATIONS

Additional books in this series can be found on Nova's website
under the Series tab.

Additional e-books in this series can be found on Nova's website
under the e-book tab.

CHEMISTRY RESEARCH AND APPLICATIONS

HAFNIUM

CHEMICAL CHARACTERISTICS, PRODUCTION AND APPLICATIONS

HONGYU YU
EDITOR

New York

NOTICE TO THE READER

The Publisher has taken reasonable care in the preparation of this book, but makes no expressed or implied warranty of any kind and assumes no responsibility for any errors or omissions. No liability is assumed for incidental or consequential damages in connection with or arising out of information contained in this book. The Publisher shall not be liable for any special, consequential, or exemplary damages resulting, in whole or in part, from the readers' use of, or reliance upon, this material. Any parts of this book based on government reports are so indicated and copyright is claimed for those parts to the extent applicable to compilations of such works.

Independent verification should be sought for any data, advice or recommendations contained in this book. In addition, no responsibility is assumed by the publisher for any injury and/or damage to persons or property arising from any methods, products, instructions, ideas or otherwise contained in this publication.

This publication is designed to provide accurate and authoritative information with regard to the subject matter covered herein. It is sold with the clear understanding that the Publisher is not engaged in rendering legal or any other professional services. If legal or any other expert assistance is required, the services of a competent person should be sought. FROM A DECLARATION OF PARTICIPANTS JOINTLY ADOPTED BY A COMMITTEE OF THE AMERICAN BAR ASSOCIATION AND A COMMITTEE OF PUBLISHERS.

Additional color graphics may be available in the e-book version of this book.

Library of Congress Cataloging-in-Publication Data

Hafnium : chemical characteristics, production. and applications / editor, HongYu Yu (South University of Science, Technology of China, Shenzhen, China).
     pages cm. -- (Chemistry research and applications)
  Includes bibliographical references and index.
  ISBN 978-1-63463-164-8 (hardcover)
  1. Hafnium. 2. Transition metals. 3. Thin films. I. Yu, Hongyu, 1976- editor.
  QD181.H5H34 2014
  661'.0514--dc23
                              2014037731

Published by Nova Science Publishers, Inc. † New York

CONTENTS

PREFACE

Hafnium is a chemical element with the symbol Hf and atomic number 72. This book brings together contributions from experts in the related fields to illustrate the significance importance of hafnium in different forms. This book is composed of 6 chapters written by experts in the respective fields making this book valuable to the group-IV materials science communities. The subjects presented in the book offers a general overview of the important issues of various hafnium compounds (e.g., Hafnium nitride, Hafnium carbide and Hafnium oxide). We believe the book is of great scientific and educational value for many readers. Readers thus can familiarize themselves with the latest knowledge in the related fields.

I am indebted to the support provided by all of those involved in the work, especially to the authors who generously agreed to share their knowledge. I would also like to thank Nova Publishers who gave me their full support to complete this book.

HongYu Yu, Ph.D.
South University of Science
Technology of China
Shenzhen, China
Tel: 0755-88018508
E-mail: yu.hy@sustc.edu.cn

In: Hafnium   ISBN: 978-1-63463-164-8
Editor: HongYu Yu   © 2015 Nova Science Publishers, Inc.

Chapter 1

A REVIEW ON HAFNIUM: SYNTHESIS, PROPERTIES AND APPLICATIONS OF VARIETY OF HAFNIUM COMPOUNDS

*Munusamy Thirumavalavan** and *Jiunn-Fwu Lee*
Graduate Institute of Environmental Engineering,
National Central University, Chung-Li, Taoyuan County, Taiwan

ABSTRACT

A series of hafnium compounds were discussed in their chapter. Synthesis, characterization and application of several hafnium compounds, hafnium complexes and hafnium nanocomposites are explored in details.

Keywords: Hafnium, compounds, properties, Applications

INTRODUCTION

Just as in real life, in the periodic table also, not all the elements play leading roles. Depending upon their aptitudes, some elements are considered to be more prominent than others [1]. For example titanium and zirconium

---

* Corresponding Author address. Email: mtvala@yahoo.com.

dominate group (IV) elements, but third member of this group hafnium appears strangely insignificant and hafnium's position as a latecomer is apparent even historically. While titanium and zirconium were discovered in the 18th century, proof of the existence of hafnium existed only in 1923 [1]. Thus, the reason for this delay in entering "the stage" of elements is particularly due to the close resemblance of both zirconium and hafnium, i.e., most of the cases hafnium was considered as zirconium only.

As a consequence of the lanthanide contraction, both elements practically possess identical metal and ionic radii due to which the presence of zirconium in hafnium minerals remained undiscovered. However, very recently researches on hafnium show that hafnium compounds can favorably compete with zirconium counterparts in various fields. A major difference between zirconium and hafnium is the formation of σ and π bonds [2,3]. Thus the method for the separation of hafnium from zirconium was clearly explained by Hevesy in chemical reviews [4]. Moreover, complexes of Hf(IV) have been examined in recent years for their potential use as ionic conductors, ferro-electrics and luminophores. Therefore, understanding the structure-property relationship for hafnium complexes has a cardinal importance in rational design of a material's properties. However, extensive studies of hafnium are sparse and thus the present review covers the significance, characteristics and application of hafnium compounds.

PROPERTIES OF HAFNIUM

So far it was chiefly ventured to elucidate the properties of the compounds of hafnium. The only statements available on the properties of metallic hafnium are the following: It has the same crystalline structure as the metallic zirconium [5]. The metal was prepared by reducing H_2HfF_6 with sodium. Taking into account that the presence of 1 percent ZrO_2 lowers the atomic weight by 1.4 units, and the atomic weight of hafnium works out to be 178.64 to 178.59 [6]. The optical spectrum of hafnium has been observed in the region between 7240.9 and 2253.95 A.U [7]. It is of interest to note that several of the strongest hafnium lines are present as weak lines in the zirconium spectrum as measured by earlier investigators [8,9]. This was because of the reason that all zirconium minerals and consequently all commercial zirconium preparations contain hafnium.

Hafnium is a hard, heavy, somewhat ductile metal having an appearance slightly darker than that of stainless steel. Hafnium's aqueous chemistry is characterized by a high degree of hydrolysis, the formation of polymeric species, a very slow approach to true equilibrium, and the multitude of complex ions that can be formed. Hafnium is a highly reactive metal. The only important oxidation number is +4. Hafnium is used in nuclear control rods, nickel-based superalloys, nozzles for plasma arc metal cutting, and high temperature ceramics. Most hafnium compounds have been of slight commercial interest aside from intermediates in the production of hafnium metal. However, hafnium oxide, hafnium carbide, and hafnium nitride are quite refractory and have received considerable interest as the most refractory compounds of the Group 4 (IVB) elements. The growing interest in the field of hafnium chemistry has arisen primarily from the tremendous importance in materials engineering and in technology. Studies on hafnium complexes showed to be attractive for an extensive range of novel molecular and supramolecular materials. Hafnium, Hf, is in group 4 (IVB) of the periodic table, as are the lighter elements zirconium and titanium. It is always found associated with the more plentiful zirconium. Zirconium and hafnium are difficult to prepare as pure metals. They are very infusible, and reach vigorously at high temperatures. The two elements are almost identical in chemical behavior. Thus formation of chelate complexes which are soluble in organic solvents is of great importance and complexes of group 4 metals have emerged as versatile catalysts for polymerization of alpha olefins. Such catalysts typically include geometrical relationship, and a chelating ligand or ligands that control the electronic and steric environments thereby affecting the activity of the catalyst and the characteristics of the resulting polymer including its molecular weight and type and degree of stereo-regularity. Unlike complexes of titanium(IV) and the group V and VI transition metals, zirconium and hafnium complexes have contributed greatly to the observed structural diversity, displaying a particularly rich and interesting stereochemistry with various coordination numbers and different structural motifs. In addition to the fundamental interest in elucidating the topological relationship between the structure of hafnium complexes and the nature of the counter-ion and reaction conditions, these compounds have many important practical applications.

Hafnium is in solid state at room temperature. It is a d-block element with atomic number 72. It's electronic configuration is [Xe] $4f^{14}5d^{2}6s^{2}$ and its relative atomic mass is 178.49. It's density is 13276 Kg m^{-3}. The melting point is 2233 °C (4051.4 °F or 2506.15 K) and boiling point is 4600 °C (8312.4 °F or 4873.15 K). The key isotopes of hafnium are ^{177}Hf, ^{178}Hf, ^{180}Hf.

APPLICATIONS AND EFFECTS OF HAFNIUM

Hafnium and its alloys are used for control rods in nuclear reactors and nuclear submarines because hafnium is excellent at absorbing neutrons, it has a very high melting point and it is corrosion resistant. It is used in high-temperature alloys and ceramics, since some of its compounds are very refractory: they will not melt except under the most extreme temperatures. Hafnium ores are rare, but two are known: hafnon and alvite. Industrial production of hafnium metal is not much more than 50 tonnes a year. Known reserves are not recorded, but can be estimated from those of zirconium. Hafnium metal does not normally cause problems but all hafnium compounds should be regarded as toxic although initial evidence would appear to suggest the danger is limited. The metal dust presents a fire and explosion hazard. Hafnium metal has no known toxicity. The metal is completely insoluble in water, saline solutions or body chemicals. Exposure to hafnium can occur through inhalation, ingestion, and eye or skin contact. Overexposure to hafnium and its compounds may cause mild irritation of the eyes, skin, and mucous membranes. No signs and symptoms of chronic exposure to hafnium have been reported in humans. Hafnium poses no threat to plants. Plants take up small amounts of hafnium from the soil in which they grow. Data on the toxicity of hafnium metal or its dust are scant. Animal studies indicate that hafnium compounds cause eye, skin, and mucous membrane irritation, and liver damage. The oral LD50 for hafnium tetrachloride in rats is 2,362 mg/kg, and the intraperitoneal LD 50 in mice for hafnium oxychloride is 112 mg/kg. (LD50 = Lethal dose 50 = Single dose of a substance that causes the death of 50% of an animal population from exposure to the substance by any route other than inhalation. LD50 is usually expressed as milligrams or grams of material per kilogram of animal weight (mg/kg or g/kg).) No negative environmental effects have been reported.

HAFNIUM COMPOUNDS

The nitrate, chloride, bromide, iodide, perchlorate and sulfate of hafnium are soluble in acid solution. Among the more insoluble compounds, phosphate is insoluble which precipitates even from 20 % sulfuric acid. The iodate precipitates from 8 M HNO_3. The hydroxide (or hydrous oxide) is precipitated with ammonia or alkali hydroxide, and a peroxide with H_2O_2 from dilute acid. Of these, only the phosphate is soluble in excess reagent. In addition to this, hafnium forms complex ions with many substances. Thus, the basic properties and applications of general hafnium compounds are already available online. Thus, here we focused on compounds/complexes of hafnium reported in recent researches.

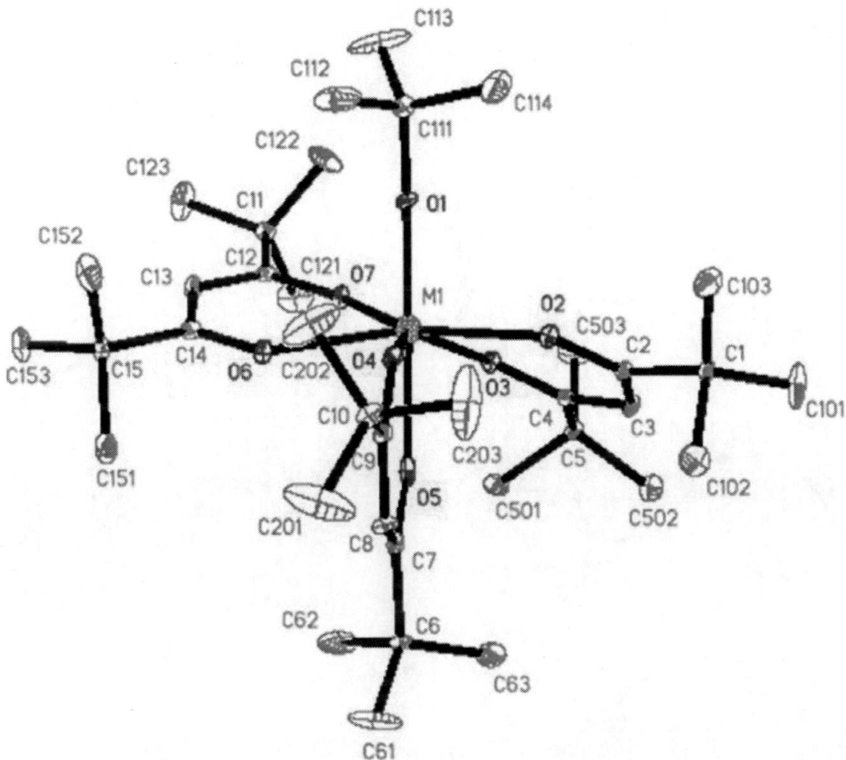

Figure 1. Molecular structure of Hf(thd)₃(OᵗBu) with 10% probability ellipsoids [14].

Hafnium Oxide

Nanostructured hafnium dioxides are requested today as materials for catalysts [10], metal oxides composites with enhanced mechanical properties [11], sensors [12] and electrolytes [13]. The approaches to such oxides preparations are often exploiting metal-organic precursors such as metal alkoxides and heteroleptic metal alkoxides and are based on soft chemistry techniques, such as metal organic chemical vapor deposition (MOCVD) and metal organic decomposition (MOD). For both approaches the choice of precursor compound is rather crucial. Thus variety of hafnium oxides such as $Hf(OtBu)_2(thd)_2$, $Hf(OtBu)(thd)_3$ and $Hf_2(OH)(O^nC5H_{11})(O^tBu)_2(thd)_4$ are reported by Spijksma et al. [14], by reacting hafnium t-butoxide with 2,2,6,6,-tetramethyl-3,5-heptanedion (Hthd). Molecular structures of $Hf(thd)_3(O^tBu)$ and $Hf_2(OH)(O^nC5H_{11})(O^tBu)_2(thd)_4$ are given in Figure 1 and Figure 2 respectively.

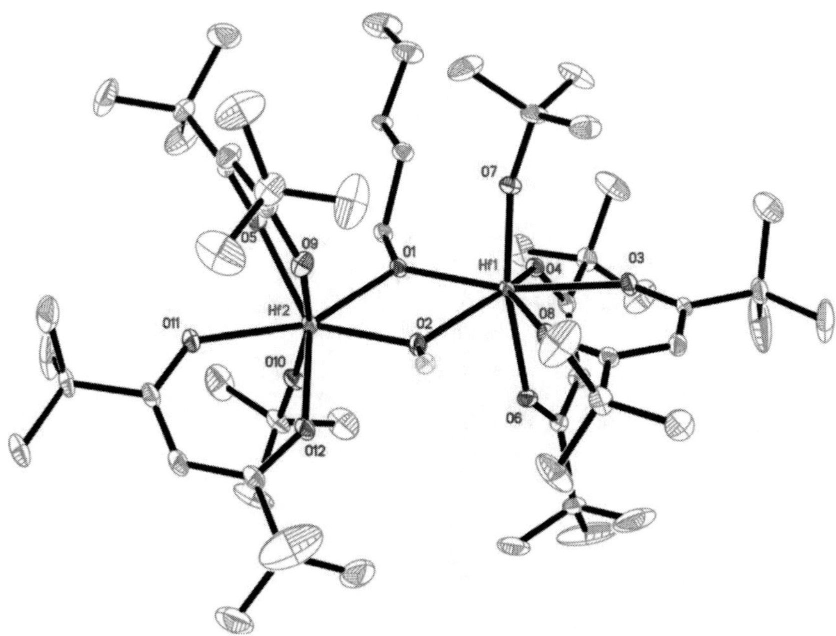

Figure 2. Molecular structure of $Hf_2(\mu\text{-}OH)(\mu\text{-}O^nC5H_{11})(O^tBu)_2(thd)_4$ with 10% probability ellipsoids [14].

They concluded that the modification of hafnium t-butoxides entails di- and tri-substituted compounds. The size of the t-butoxide ligands played a major role to form di-substituted compounds instead of the mono-substituted propoxide compounds. The metal-oxygen bond length of the t-butoxide ligands is significantly longer compared to that of propoxide in analogous compounds. This clearly highlights the presence of sterical hindrance in these systems. The stability of the formed compounds is tremendously high, i.e., upon storage for half a year, no change in composition was observed.

HfO_2 thin films (80 nm thick) were fabricated and reported by Ramzan et al., [15] using electron beam evaporation technique at various substrate temperatures ranging from 25 to 120 °C. These films were then thermally annealed at 500 °C for one and half hour in vacuum. After thermal annealing, films were characterized by XRD, AFM and spectrophotometer. In this regard, it was observed that the as-deposited HfO_2 films were mostly amorphous in nature and transformed to polycrystalline with monoclinic structure after annealing at 500 °C. Moreover, films fabricated at different substrate temperatures revealed different morphologies and crystallite orientations on thermal annealing. Such different morphologies and crystallite orientations appear to be responsible for any variations in the surface roughness and the optical properties e.g., optical band gap energy (3.4-3.65 eV), refractive index (1.25-2.55), extinction coefficient (0.25-0.46) etc. These optical properties demonstrate oscillatory behavior with different substrate temperatures due to crystallite growth along different preferred orientations. On the basis of above mentioned facts, it can be concluded that the post thermal annealing demonstrates better tendency to change the structural and optical properties of HfO_2 thin films. In addition, annealed HfO_2 films showed better reflectivity (5-10%) in the NIR region which can further be improved by inserting a metallic layer into the oxide-metal-oxide (O-M-O) structure. Hence, such O-M-O structures can be useful for heat mirror applications. As shown in Figure 3, only a broad peak was observed for the as deposited thin films which confirmed their amorphous nature.

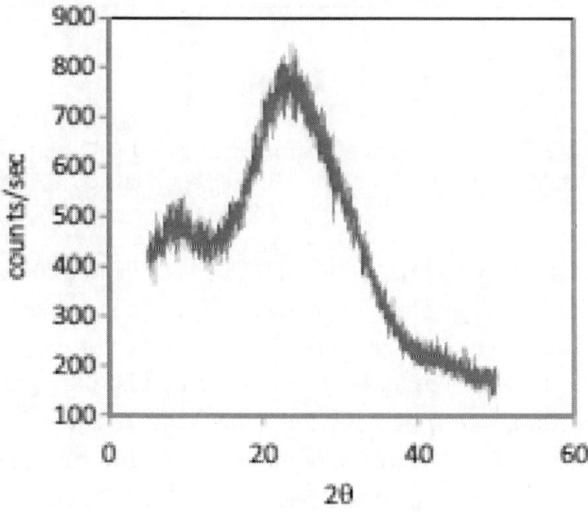

Figure 3. X-ray diffraction patterns of as synthesized HfO_2 thin films before thermal annealing [15].

The XRD studies were also carried out for thermally annealed HfO_2 films deposited at various substrate temperatures, as shown in Figure 4, which revealed that thermal annealing has made structural changes from amorphous to poly crystalline state.

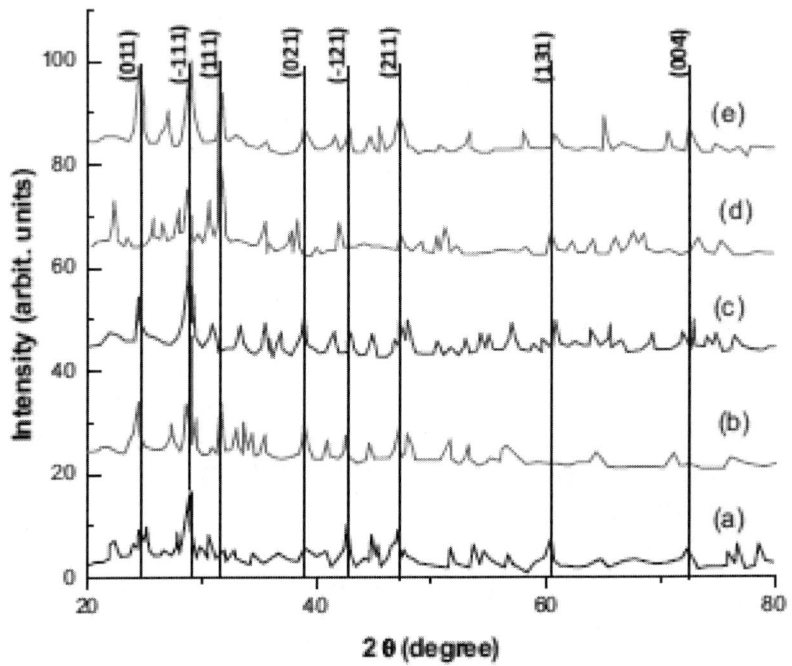

Figure 4. X-ray diffraction spectra of thermally annealed HfO_2 thin films deposited at various substrate temperatures (a) 25 °C; (b) 50 °C, (c) 80 °C, (d) 100 °C and (e) 120 °C [15].

Also in 2010, Mendoza et al., [16] have reported the preparation of hafnium oxide (HfO_2) films deposited by the ultrasonic spray pyrolysis process. The films were synthesized from hafnium chloride as raw material in deionized water as solvent and were deposited on corning glass substrates at temperatures from 300 to 600 °C. For substrate temperatures lower than 400 °C, the deposited films were amorphous, while for substrate temperatures higher than 450 °C, the monoclinic phase of HfO_2 appeared. Scanning electron microscopy showed that the film's surface was rough with semi-spherical promontories. The films showed a chemical composition close to HfO_2, with an Hf/O ratio of about 0.5. UV radiation was used in order to achieve the thermoluminescent characterization of the films; the 240 nm wavelength induced the best response. In addition, preliminary photoluminescence (PL) spectra, as a function of the deposition temperatures, are shown. PL emission spectra, as a function of the substrate temperature (300, 400, 500, 600 °C), for hafnium oxide films are shown in Figure 5. Here it is possible to distinguish three emission bands centered at 425, 512, and 650 nm when excited with a 254 nm wavelength. The results presented in this figure showed that the amorphous films grown at 300 and 400 °C possess different photoluminescence characteristics compared with the crystalline samples containing monoclinic phase, 500 and 600 °C. The samples deposited at low substrate temperatures clearly showed the band centered at 425 nm; the other bands (512 and 650 nm) appeared in samples deposited at high substrate temperatures (500 and 600 1C). In the sample deposited at 600 °C, the band centered at 425 nm has a higher intensity.

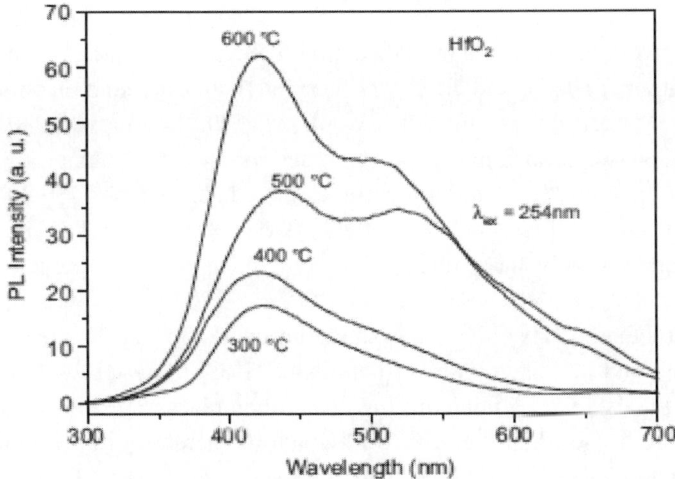

Figure 5. PL emission spectra for HfO$_2$ films deposited at 300, 400, 500, and 600 °C and excited with radiation of 254 nm [16].

Hafnium Carbide

In 2013, Zeng et al., [17] have performed a search for stable compounds in the hafnium-carbon (Hf-C) system at ambient pressure using a variable-composition *ab initio* evolutionary algorithm implemented in the USPEX code. In addition to the well-known HfC, they predicted two additional thermodynamically stable compounds Hf_3C_2 and Hf_6C_5. The structure of Hf_6C_5 with space group $C2/m$ contains 22 atoms in the conventional cell, and this prediction revives the earlier proposal by Gusev and Rempel [18]. The stable structure of Hf_3C_2 also has space group $C2/m$ and is more energetically favorable than the *Immm, P3m1, P2*, and $C222_1$ structures [18]. The dynamical and mechanical stabilities of the newly predicted structures have been verified by calculations of their phonons and elastic constants. Structural vacancies are found in the ordered defective rock-salt-type HfC. Chemical bonding, band structure, and Bader charges are presented and discussed. All three compounds are weak metals with increasing metallicity as the vacancy concentration increases. The mechanical properties of the hafnium carbides nonlinearly decrease with increasing vacancy concentration, indicating the defect tolerance of this refractory compound. It is, therefore, possible to tune the hardness, ductility, and electrical conductivity by varying the stoichiometry of the hafnium carbides. In addition to rock-salt-type HfC, they found two other compounds Hf_3C_2 and Hf_6C_5, both belonging to space group $C2/m$. The enthalpies of formation of the predicted structures are explained. It can clearly be seen that HfC, Hf_6C_5, and Hf_3C_2 are thermodynamically stable compounds. Hf_2C $(R3m)$ is very close to the convex hull curve but lies above it, i.e., it is a metastable phase. Transition-metal carbides often have numerous stable phases. For example, five titanium carbides have been reported [19,20] in the literature at ambient pressure, Ti_2C, Ti_3C_2, Ti_8C_5, Ti_6C_5, and TiC. However, hafnium carbides only have three thermodynamically stable polymorphs at ambient pressure, i.e., Hf_3C_2, Hf_6C_5 and HfC.

The structure of Hf_3C_2 has the space group $C2/m$ and 20 atoms in the conventional unit cell. Two more structures of Hf_3C_2 proposed by Gusev and Rempel [18] also were found during the structure search. However, their enthalpies are higher than that of $C2/m$, which is, therefore, more stable. The crystal structure of Hf_6C_5 showed that its space group is also $C2/m$ with 22 atoms in the conventional unit cell. All stable hafnium carbides are strongly related structures and can be derived from the cubic rock-salt-type structure of HfC. HfC has a structure of cubic-packing hafnium atoms, and carbon atoms fill all octahedral voids, which is an ideal cubic rock-salt-type structure. The octahedra shown by the green color are empty, i.e., formed by six Hf atoms but without interstitial C atoms. In the Hf_3C_2 structure, only 2/3 of the carbon octahedral voids are filled (and 1/3 are vacant), and in Hf_6C_5, 5/6 are filled (and 1/6 are vacant). Both Hf_3C_2 and Hf_6C_5 structures are carbon-deficient ordered crystals in which the vacancies appear in every second octahedral layer with 1/3 of the in-layer octahedra occupied (Hf_3C_2) or 2/3 of the in-layer octahedra occupied (Hf_6C_5). Ordering of the vacancies in both cases lowers the symmetry from cubic to monoclinic. Moreover, due to the vacancies, the coordination number of the Hf atoms varies in different systems: 6 in HfC, 5 in Hf_6C_5, and 4 in Hf_3C_2, whereas, in all these structures, carbon atoms invariably had the coordination number 6 (octahedral coordination). In this way, Hf_6C_5 and Hf_3C_2 can also be described as defective rock-salt-type structures.

HAFNIUM COMPLEXES

Significant success has been achieved in recent years in chemistry and structure of metal fluoride complexes, in particular hafnium complexes [21]. Moreover, various fluoride complexes of Hf(IV) have been examined in recent years for their potential use as ionic conductors [22], ferro-electrics [23] and luminophores [24,25]. Therefore, understanding the structure-property relationship for hafnium fluoride complexes has a cardinal importance in rational design of a material's properties. Davidovich et al., [21] in their review atcile, published reviews dedicated to the stereochemistry of fluoride complexes and mixed-ligand fluoride complexes of hafnium. The structure of complex anions (complexes) of structurally studied fluoride and mixed-ligand fluoride complexes of hafnium and the dependence of the structural motif of the complex anion (complex) on the F(L):Zr(Hf) ratio in the compound.

Tomachynski et al., [26] have reported a series of new hafnium(IV) phthalocyanines with various β-dicarbonyl ligands prepared via direct interaction between di(chloro) hafnium(IV) phthalocyanines and free β–diketones and also with 4-benzoyl-3-methyl 1-phcnyl-2-pyrazolin-5-one. The structure of the obtained bis(β-dicarbonilato) hafnium(IV) phthalocyanines was studied by two dimension ^1H NMR spectroscopy. Absorption and fluorescence spectroscopic studies have been investigated in various solvents. They found that the interaction between phthalocyanine and out of plane ligand protons allows determining the structure of macromolecule by NOESY and ROESY techniques. It was concluded that additionally the molecule of DMF is solvated in a cavity between the "out of plane" dicarbonilato ligand and the N8 macrocycle. Absorption and fluorescence spectroscopic properties in the concentration range of 10^{-6} M/dm3, indicated that there is rather no tendency to the aggregation; however, absorbance slightly decreases with higher solvent polarity.

Spijksma et al., [14] have reported the synthesis and properties of hafnium tert-butoxides and tert-butoxo-β-diketonate complexes as shown in Figure 1.

Symmetric oxamide complexes of Hf was derived from 1,2-bis(3,5-di-tert-butyl-2-hydroxiphenyl)oxamide and Hf its application as catalyst for ethylene polymerization was reported by Rodríguez et al., [27]. The syntheses were carried out using various bases and solvents (triethylamine/toluene, NaH/THF and NaOH/methanol). They have observed that these complexes have a planar structure and the hafnium derivative was obtained by means of NaH in THF. Finally, a study on the polymerization of ethylene showed that the oxamide series were active catalysts, presenting the highest activity of the series and a marked solvent effect. Thus the oxamide based hafnium complexes can act as new catalysts with potential applications in low olefin polymerization reactions.

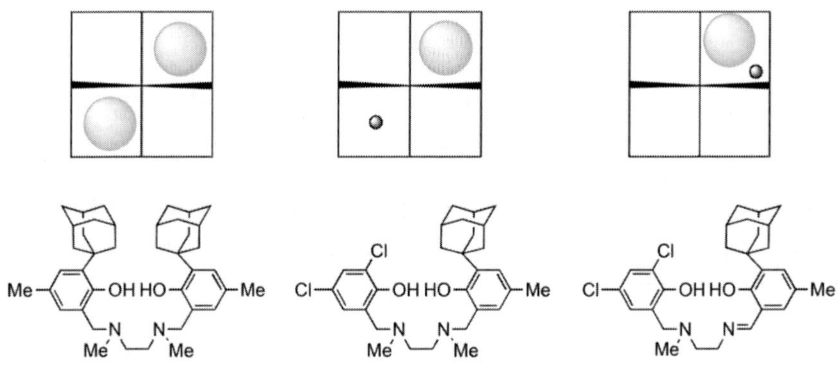

Figure 6. Schematic representation of quadrant occupancy formed by a fac–fac wrapping symmetric Salalen ligand (left), a fac–fac wrapping non-symmetric Salalenn ligand (middle), and a fac–mer wrapping Salalen ligand (right). Light spheres represent bulky alkyl groups; dark spheres represent halo groups. The monomer approaches the centrally located metal in a horizontal fashion [28].

The activity of dibenzylhafnium Salalen complexes as a catalyst in polymerisation of propylene was described by Press et al., [28]. The Salalen ligand precursors combining a bulky alkyl group (1-adamantyl) on the imine-side phenol and electron withdrawing halo groups of different sizes on the amine-side phenol were explored as shown in Figure 6. All metal complexes were obtained as single diastereomers. An X-ray crystallographic structure of a hafnium complex of an additional ligand carrying the combination of tertbutyl and chloro substituted phenolates, 4-Hf, revealed a fac–mer wrapping of the Salalen ligand around the metal centre. All complexes led to active catalysts in propylene polymerisation and to isotactic polypropylene of high regioregularity.

The Salalen ligands introduced in this work which combine an adamantyl substituted phenolate and ortho, para-halo substituted phenolate were found to wrap in a diastereoselective fac–mer manner around octahedral hafnium centres. All complexes led to active catalysts for propylene polymerization yielding polypropylene featuring a degree of isotacticity which depended both on the metal and on the size of the halo group.

The reactivity and thermal stability of hafnium complexes containing the N-alkyl-substituted amine biphenolate ligands of the type $[RN(CH_2\text{-}2\text{-}O\text{-}3,5\text{-}C_6H_2(tBu)_2)_2]^{2-}$ ($[R\text{-}ONO]^{2-}$; R = tBu, iPr, or nPr was investigated by Liang et al., [29]. Controlled experiments revealed that the production of hafnium complexes proceeds via unexpected thermal degradation that produces a highly reactive, transient ortho-quinone methide intermediate. Similar reactions, however, led to the formation of homoleptic bis-ligand complexes as colorless crystals. Decisive factors governing these divergent reaction pathways and complex constitutions are discussed. These findings are informative and relevant to the structural and reaction chemistry of complexes containing the conceivably ubiquitous Mannich-type ligands.

Controlled isotopic polymerization of α–olefins by hafnium complex incorporating with a trans-cyclooctanediyl-bridged type bis(phenolate) ligand was reported by Nakata et al., [30]. Treatment of trans-cyclooctanediyl-bridged type ligand with $Hf(CH_2Ph)_4$ in toluene afforded the corresponding hafnium(IV) dibenzyl complex as pale yellow crystals. X-ray crystallographic analysis revealed that the six-coordinated hafnium center incorporated in the ligand adopted cis-α configuration, and two benzyl ligands were coordinated to hafnium center by $\eta1$-fashion. In the polymerization of 1-hexene, the combination of hafnium complex and $B(C_6F_5)_3$, $(Ph_3C)[B(C_6F_5)_4]$, or dMAO (dried methylaluminoxane) as an activator provided poly(1-hexene)s with perfect isotacticity as shown in Figure 7. These catalytic characteristics are an improvement over other postmetallocene catalysts for α-olefin polymerization, making them an attractive option as the foundation for the development of general-purpose, new-generation industrial postmetallocene catalysts.

Tonzetich et al., [31] have reported the synthesis, characterization and activation of hafnium dialkyl complexes that contain a $C2$-symmetric diaminobinaphthyl dipyridine ligand. The diamine *rac*-H_2[MepyN] (*rac-N,N'*-di(6-methylpyridin-2-yl)-2,2' diaminobinaphthalene) has been prepared in good yield through reductive amination of *rac*-2,2'-diaminobinaphthalene with 6-methylpyridinecarboxaldehyde. Activation of [MepyN]Hf(CH$_2$CHMe$_2$)$_2$ with various Lewis acids leads to observable cationic alkyls that are not active toward 1-hexene polymerization. The chemistry of six-coordinate hafnium complexes of the [MepyN]$^{2-}$ ligand offer an opportunity to examine the reactivity of coordinatively more saturated species with well-defined activators such as boranes, carbocations, and proton sources. Activation of dialkyl species, especially [MepyN]M(CH$_2$CHMe$_2$)$_2$, gives rise to asymmetric, cationic, monoalkyl complexes that are readily observable by standard NMR techniques. However, these species will not polymerize 1-hexene. The absence of catalytic activity with these complexes can be ascribed to both their high coordination number and the presence of a sterically encumbering diamidodipyridine ligand which serves to both decrease the Lewis acidity at the metal and hinder olefin approach and binding.

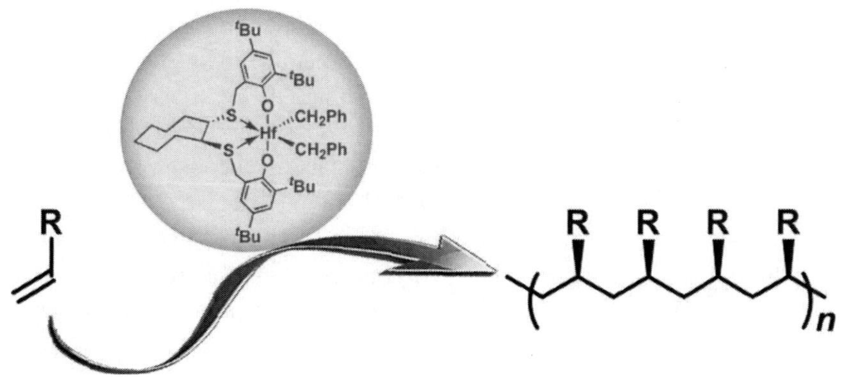

Figure 7. Polymerization of 1-hexene by hafnium complex [30].

Thompson et al., [32] have reported the synthesis, characterization and reactivity of the hafnium complex stabilized by amidate ligands. A photo and thermally stable bis(amidate)–dibenzyl complex of Hf ([^{DMP}(NO)^{Ph}]$_2$Hf(CH$_2$Ph)$_2$ (THF) was formed as a monosolvated THF adduct in near quantitative yield from Hf(CH$_2$Ph)$_4$ and *N*-2,6-dimethyl-phenyl(phenyl)amide. Isomerization between the THF-bound product and the THF-free product can be observed visually by the conversion from a red-orange product at low temperatures to a pale yellow product at high temperatures. Solid-state crystallographic characterization of the orange product confirmed its constitution as a monosolvated species. Kinetic parameters for the exchange of the THF moiety were determined from variable-temperature NMR experiments. The product of the hydrolysis of the Hf dibenzyl species by water was characterized by X-ray crystallography, and was found to be a rare tetrametallic Hf oxo cluster species.

Synthesis of Hf(IV) complex from cyclam (1,4,8,11-tetraazacyclotetradecane)-containing amino-amido groups and its application as a catalyst for direct esterification of 4-phenylbutyric acid with benzyl alcohol was reported by Thirumavalvan and Lee [33]. In this case, cyclam has to undergo structural transformation to bind the four nitrogen atoms to the same metal center. This complex exemplifies a direction in the use of amido-amino ligands and in particular in the form of the macrocycle.

HAFNIUM NANOCOMPOSITES

In recent years, a great deal of attention has been paid to various approaches to produce metal-polymer nanocomposites containing ultrafine particles of metals or their compounds in polymer matrices through polymer synthesis. The development of effective processes for the fabrication of high-temperature composites based on refractory compounds, such as hafnium (Hf) oxides and carbides, is of considerable interest for the preparation of thermally stable, high-strength coatings. Hafnium compounds are widely used in the fabrication of composites capable of withstanding high temperatures and aggressive attack and in basic organic synthesis, in particular as Ziegler-Natta catalysts.

Pomogailo et al., [34] have reported the hafnium containing nanocomposites using polymer mediated synthesis and controlled thermolysis. The new types of Hf-containing monomers and polymers were synthesized. The composition and structure of the synthesized hafnium-containing precursors and thermolysis products are determined by elemental analysis, IR spectroscopy, mass spectrometry, optical and scanning electron microscopy, and x-ray diffraction. Thermodynamic analysis is used to assess the equilibrium composition of the Hf-C-H-O system and establish the conditions under which HfC and HfO_2 are formed. Polymers containing Hf(IV) in the main chain or side chains can be prepared by any of the known processes for synthesis of metal-containing polymers. An attractive approach is the polycondensation of dicyclopentadienyl derivatives of hafnium(IV) with diols or oxygen-free diacetylenic ligands. An alternative approach is polymerization of the Hf(IV) monomers with (meth)acrylate and fumarate groups, synthesized for the first time in this study. The thermolysis of the synthesized organohafnium polymers was investigated. Compositions of the gas and solid phases that were formed in the static nonisothermal system were determined. This analysis, in conjunction with X-ray diffraction results, indicates that pyrolysis leads to the formation of metal-polymer nanocomposites consisting of crystalline HfO_2 and HfC stabilized by the polymer matrix. The hafnium carbide content depends on the ligand shell of Hf(IV) and thermolysis conditions. Kinetically, the polymer-mediated synthesis of nanoparticles is a two-step process involving polymerization transformations of the metal-containing precursor, which initiate subsequent transformations of the polymer formed. In this process, the free energy liberated during thermolysis exceeds that absorbed by the forming metal-containing polymer. In other words, the free energy liberated in chemical reactions is consumed by other, concurrent processes. Solid-state reactions, including thermolysis, can be initiated by transformations taking place around structural inhomogeneities and propagating to structurally homogeneous regions. The authors hope that further research will make it possible to optimize both the synthesis and thermolysis of Hf-containing polymers for the preparation of HfC-based nanocomposites.

Dahal and Chikan [35] have developed a cheap composite hafnium oxide-gold core-shell nanoparticle system as shown in Figure 8 that combines a high dielectric constant with good conductivity is important for the future of the electronic industry. In this study, two different sizes, 7.3 ± 2.2 and 5.6 ± 1.9 nm, of HfO_2@Au core-hell nanoparticles are prepared by using a high-temperature reduction method. The core-shell nanoparticles are characterized by powder X-ray diffraction, high-resolution transmission electron microscopy (HRTEM), energy dispersive X-ray analysis (EDX), and UV-visible absorption spectroscopy. HfO_2 exhibits no absorption in the visible region, but the HfO_2@Au core-shell nanoparticles show a plasmon absorption band at 555 nm that is 25 nm red-shifted as compared to pure gold nanoparticles as shown in Figure 9. According to transmission electron microscopy and energy dispersive X-ray analysis, the HfO_2 particles are coated with approximately three atomic layers of gold.

Thus they have concluded that HfO_2@Au core-shell nanoparticles are successfully prepared by a high-temperature reduction method in solution. The growth of such structures is based on the seeded nucleation of gold on the surface of hafnium oxide. High-quality, HfO_2@Au can be prepared by a quick and simple solution processing and be easily modified.

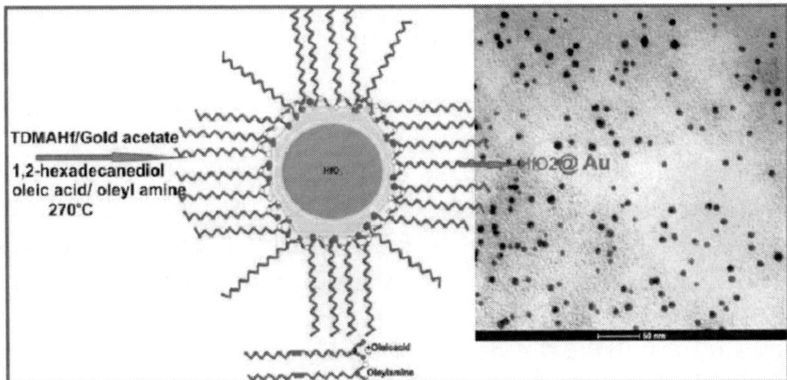

Figure 8. Hafnium oxide-gold core-shell nanoparticle prepared by Dahal and Chikan [35].

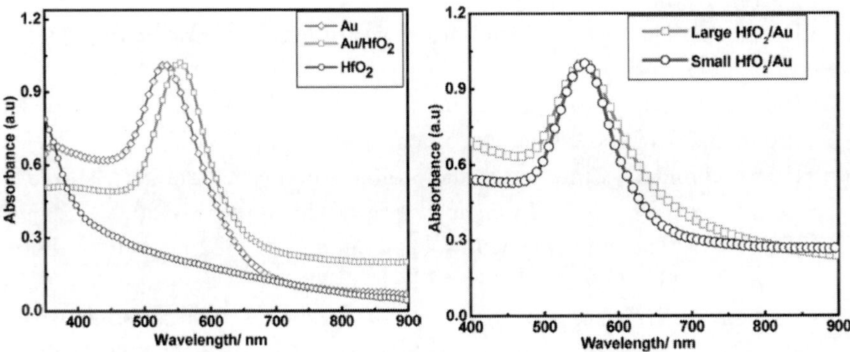

Figure 9. (a) UV-visible absorption spectra of pure gold nanoparticles (diamond curve), $HfO_2@Au$ core-shell nanoparticles (square curve), and bare hafnium oxide nanoparticles (circle curve); (b) UV-visible absorption spectra of two different sizes of hafnium oxide-gold core-shell nanoparticles [35].

CONCLUSION

This chapter covers the synthesis and application of variety of hafnium with different formation mechanism. The several synthetic approaches and applications explained in this chapter clearly emphasized the significance importance of hafnium in different possible forms.

REFERENCES

[1]   Marschner C. Hafnium: Stepping into the limelight. *Angewandte Chemie International Education*. 2007; 46: 6770-6771.

[2]   Krueger C, Mueller G, Erker G, Dorf. U, Engel K. Structural and chemical features of early transition metal compounds: Notable differences between corresponding pairs of (s-cis-.η^4-conjugated diene)zirconocene and-hafnocene complexes. *Organometallics*. 1985; 4: 215-223.

[3]   Jacobsen H, Berke H, Brackemeyer T, Eisenblätter T, Erker G, Fröhlich R, Meyer O, Bergander K. Compariosn of characteristic structural features among the triade of tris(cyclopentadienyl)(group-4 metal) complex cations: a combined theoretical and experimental study. *Helvetica Chimica Acta*. 1998; 81: 1692-1709.

[4]   Hevesy G. The discovery and properties of hafnium. *Chemical Reviews*. 1925; 2(1): 1-41.

[5]   Welsbach AO. *On the elements of the ytterbium group*. Berichte der Akademie der Wissenschaften, Wien. 1906; 115: 737-747.

[6]   Hönigschmid O, Zintl E. On the atomic weight of hafnium. *Zeitschrift fur anorganische und allgemene chemie*. 1924; 140: 335-336.

[7]   Hansen W. The optical spectrum of hafnium; On Urbain's celtium lines. *Nature*. 1923; 111: 322- 461.

[8]   Bachem B. *"The optical spectrum of zirconium"*. Dissertation Bonn. 1910.

[9]   Exner H. *"Die Spektren der Elemente"*. 1911; 11: 38-49.

[10]   Magrez A, Smajda R Seo JW, Horvath E, Ribic PR, Andresen JC, Acquaviva D, Olariu Λ, Laurenczy G, Forro L. Striking influence of the catalyst support and its acid-base properties: New insights into the growth mechanisms of carbon nanotubes. *ACS Nano*. 2011; 5: 3428-3437.

[11]   Chang CI, Wang YN, Pei HR, Lee CJ, Du XH, Huang JC. Superplastic AZ31 Mg tubes for hydroforming. *Key Engineering Materials*. 2007; 351: 114-119.

[12]   Wu YZ, Hu FT, Gan N, Hou JG, Li TH, Cao YT, Zheng L. Zirconium and hafnium tert-butoxides nanomaterials. *Advanced Materials Research*. 2011; 490: 343-344.

[13]   Chao CC, Hsu CM, Cui Y, Prinz FB. Improved solid oxide fuel cell performance with nanostructured electrolytes. *ACS Nano*, 2011; 5: 5692-5696.

[14] Spijksma GI, Seisenbaeva GA, Bouwmeester HJM, Blank DHA, Kessler VG, Zirconium and hafnium tert-butoxides and tert-butoxo-β-diketonate complexes - Isolation, structural characterization and application in the one-step synthesis of 3D metal oxide nanostructures. *Polyhedron*. 2013; 53: 150-156.

[15] Ramzan M, Wasiq MF, Rana AM, Ali S, Nadeem MY. Characterization of e-beam evaporated hafnium oxide thin films on post thermal annealing. *Applied Surface Science*. 2013; 283: 617-622.

[16] Mendoza JG, Frutis MAA, Flores GA, Hipólito MG, Cerda AM, Nieto JA, Montalvo TR, Falcony C. Synthesis and characterization of hafnium oxide films for thermo and photoluminescence applications. *Applied Radiation and Isotopes*. 2010; 68: 696-699.

[17] Zeng Q, Peng J, Oganov AR, Zhu Q, Xie C, Zhang X, Dong D, Zhang L, Cheng L. Prediction of stable hafnium carbides: their stoichiometries, mechanical properties, and electronic structure. *Physical Review B*. 2013; 88: (214107-1)-(214107-6).

[18] Gusev AI, Rempel AA. Superstructures of non-stoichiometric interstitial compounds and the distribution functions of interstitial atoms. *Physica status solidi* (a). 1993; 135: 15-58.

[19] Jain A, Ong SP, Hautier G, Chen W, Richards WD, Dacek S, Cholia S, Gunter D, Skinner, D, Ceder G, Persson KA. Commentary: The materials project: A materials genome approach to accelerating materials innovation. APL Materials. 2013; 1: (011002-1)- (011002-11).

[20] Dzhalabadze NV, Eristavi BG, Maisuradze NI, Kuteliya E. Structure of crystalline phases in Ti-C thin films. *The Physics of Metals and Metallography*. 1998; 86: 59-64.

[21] Davidovich RL, Marinin DV, Stavila V, Whitmire VH. Stereochemistry of fluoride and mixed-ligand fluoride complexes of zirconium and hafnium. *Coordination Chemistry Reviews*. 2013; 257: 3074-3088.

[22] Mallikarjunaiah KJ, Ramesh KP, Damle R. ^{1}H and ^{19}F NMR relaxation time studies in $(NH_4)_2ZrF_6$ superionic conductor. *Applied Magnetic Resonance*. 2009; 35: 449-458.

[23] Bauer MR, Pugmire DL, Paulsen BL, Christie RJ, Arbogast DJ, Gallagher CS, Raveane EV, Nielson RM, Ross II CR., Photinos P, Abrahams SC. Aminoguanidinium hexafluorozirconate: a new ferroelectric. *Journal of Applied Crystallography*. 2001; 34: 47-54.

[24] Godneva MM, Motov DL, Boroznovskaya NN, Klimkin VM. Synthesis of zirconium (hafnium) fl**oride complexes and their X-rau

luminescence properties. *Russian Journal of Inorganic Chemistry.* 2007; 52: 661-666.

[25] Underwood CC, McMillen CD, Chen H, Anker JN, Kolis JW. Hydrothermal chemistry, structures, and luminescence studies of alkali hafnium products. *Inorganic Chemistry.* 2013; 52: 237-244.

[26] Tomachynski LA, Tretyakova IN, Chernii VY, Volkov SV, Kowalska M, Legendziewicz J, Gerasymchuk YS, Radzki S. Synthesis and spectral properties of Zr(IV) Hf(IV) pthalocyanines with β-diketonates as axial ligands. *Inorganica Chimica Acta,* 2008; 361: 2569-2581.

[27] Rodríguez MG, -Pérez VMJ, -Rivera JEH, Domínguez JM, Contreras R, Quijada R. Synthesis, characterization and ethylene polymerization activity of titanium, zirconium and hafnium compounds derivatives from symmetric oxamides. *Polyhedron.* 2007; 26: 4321-4327.

[28] Press K, Venditto V, Goldberg I, Kol M. Zirconium and hafnium salalen complexes in isospecific polymerization of propylene. *Dalton Transactions. 2013*; 42: 9096-9103.

[29] Liang L-C, Chien C-C, Chen M-T, Lin S-T. Zironium and hafnium complexes containing N-alkyl-substituted amine biphenolate ligands: Unexpected ligand degradation and divergent complex constitutions governed by N-alkyls. *Inorganic Chemistry.* 2013; 52: 7709-7716.

[30] Nakata N, Toda T, Matsuo T, Ishii A. Controlled isospecific polymerization of α-olefins by hafnium complex incorporating with a tans-cyclooctanediyl-bridged [OSSO] type bis (phenolate) ligand. *Macromolecules.* 2013; 46: 6758-6764.

[31] Tonzetich ZJ, Schrock RR, Hock AS, Müller P. Synthesis, characterization, and activation of ziroconium and hafnium dialkyl complexes that contain a C_2-symmetic diaminobinaphthyl dipyridine ligand. *Organometallics.* 2005; 24: 3335-3342.

[32] Thomson RK, Patrick BO, Schafer LL. Synthesis, characterization and reactivity of the first hafnium alkyl complex stabilized by amidate ligands. *Canadian Journal of Chemistry.* 2005; 83: 1037-1042.

[33] Thirumavalavan M, Lee J-F. Synthesis of Hf(IV) complex from cyclam (1,4,8,11-tetraazacyclotetradecane)-containing amino-amido groups. *Synthetic Communications.* 2012; 42: 223-226.

[34] Pomogailo AD, Dzhardimalieva GI, Rozenberg AS, Kestelman VN. Hafnium-containing nanocomposites. *Journal of Thermoplastic Composite Materials.* 2007; 20: 151-174.

[35] Dahal N, Chikan V. Synthesis of hafnium oxide-gold core-shell nanaoparticles. *Inorganic Chemistry.* 2012; 51: 518-522.

In: Hafnium
Editor: HongYu Yu

ISBN: 978-1-63463-164-8
© 2015 Nova Science Publishers, Inc.

Chapter 2

MECHANICAL ENGINEERING OF HAFNIUM WITH METAL TRANSITION MULTILAYERS

Cesar Escobar[1], Julio C. Caicedo[1,2]*
and William Aperador[3]

[1]Thin-Film Group, Universidad del Valle – Cali, Colombia
[2]Tribology Polymers, Powder Metallurgy and Processing of Solid
Recycled Research Group Universidad del Valle, Cali, Colombia
[3]Facultad de Ingeniería, Universidad Militar Nueva Granda,
Bogotá, Colombia

ABSTRACT

Mechanical and tribological evolution on 4140 steel surfaces coated with Hafnium with Metal transition Multilayers systems deposited in various bilayer periods (Λ) via magnetron sputtering has been exhaustively studied in this work. The coatings were characterized in terms of structural, chemical, morphological, mechanical and tribological properties by X-ray diffraction (XRD), X-ray photo electron spectroscopy (XPS), atomic force microscopy, scanning and transmission electron microscopy, nanoindentation, pin-on-disc and scratch tests. The failure mode mechanisms were observed via scanning electron microscopy. X-ray diffraction results showed preferential growth in the face-centered cubic (111) crystal structure for Hafnium with Metal transition

---

* Corresponding author address: Email: jcaicedoangulo1@gmail.com. Tel.: +57 2 339 46 10; fax: +57 2 339 32 37.

Multilayers as hard coatings. The best enhancement of the mechanical behavior was obtained when the bilayer period (Λ) 15 nm (n = 80), yielding the highest hardness, highest elastic modulus, the lowest friction coefficient and the highest critical load. These results indicate enhancement of mechanical, tribological, and adhesion properties, compared to HfN/VN multilayer systems with bilayer period (Λ) of 1200 nm (n = 1). So, these multilayered are present as coatings to surface on tool machining with excellent industrial applications.

Keywords: Hard coatings, quasi-isostructural system, superlattices, tool wear

INTRODUCTION

Scientific literature over the last years have shown that multilayered films of transition metal nitrides have received sizeable interest for their outstanding properties within a narrow range of bilayer thicknesses [1]. Superior mechanical properties like extreme hardness and good wear resistance have been reported [2, 3]. Therefore, during recent years, transition metal nitride materials have developed crucial roles in applications where these properties are required. Control of the microstructural and surface morphological evolution of polycrystalline transition metal nitride films is essential for the performance and lifetime of coated tools. For TiN/VN, TiN/NbN, and W/WN the transition metal nitrides have been studied for their excellent mechanical properties [4, 5]. Growth of such multilayered structures with nanometer-scale dimensions has attracted much attention from the scientific and industrial community. These coatings, now called heterostructures, such as alternate a-Si3N4/nc-TiN, or TiAlN/SiNx coatings exhibit relative high hardness (~40 GPa) and offer potential advantages for dry milling, drilling, and turning[6, 7]. For other bilayer combinations (CrN/WN, and TiN/NbN) maximum hardness values do not exceed 30 GPa, but they are still interesting for mechanical applications; moreover, new tribological and physicochemical properties like very low friction coefficient and anticorrosive properties are now widely studied [7–9]. However, very few reports persist in the literature about mechanical and tribological responses in nanostructured systems based on isostructural assembly from nitride coatings generated by transition metals.

This work evaluated the influence of Hafnium with Metal transition Multilayers deposited onto silicon (100) and industrial steel substrates with different bilayer periods (Λ) and bilayer numbers (n) on their physical nature compared to HfN, VN single layered coatings and uncoated industrial

AISI4140 steel. Novel results are shown in this paper for possible surface applications in industrial factories under processes with aggressive environments, as in the metal used in the mechanical industry. For this purpose, silicon and AISI4140 steel substrates were coated with a set of HfN/VN superlattices with Λ between 1200 and 15 nm and n between 1 and 80, with a total thickness of 1.2 μm.

EXPERIMENTAL DETAILS OF MECHANICAL ENGINEERING OF HAFNIUM WITH METAL TRANSITION MULTILAYERS

Coating Deposition

In this work, Hafnium with Metal transition Multilayers were deposited onto Si (100) and AISI4140 steel substrates by using a multi-target r.f. magnetron sputtering system, with an r.f. source (13.56 MHz) for the applied negative voltage bias on the substrate and two Hf and V 10 cm diameter targets with 99.9% purity. A 350-Wmagnetron power was applied to the hafnium target, while a power of 400W was applied to the vanadium target. The deposition chamber was initially pumped down to less than 5×10^{-6} mbar by using a turbo molecular pump and then a mixture of (80%) for Ar gas and (20%) for N2 gas was introduced into the chamber. An r.f. negative bias voltage of -30 V was used; substrate temperature was around 250 °C and a substrate-to-target distance of 7 cm. for both coatings. During growth, the chamber pressure was maintained at 2×10^{-3} mbar. For multilayered depositions, the hafnium and vanadium targets were covered periodically with a steel shutter. Before deposition, the targets and substrates were sputter-cleaned during a 20-min period. An exhaustive XRD study was carried out by using a PANalytical X'Pert PRO diffractometer with Cu-Kα radiation ($\lambda =$ 15.406 nm) at Bragg–Brentano configuration ($\theta/2\theta$) in high- and low-angle ranges. Bilayer periods in multilayers were measured by using low-angle XRD $\theta/2\theta$ scans and were compared to those obtained from transmission electron microscopy (TEM) micrographs. Micro structural analysis of multilayers was mainly performed by TEM, using a Philips CM 30 microscope operating at 300kV. The ratio between thicknesses of HfN and VN single layers wasobtained by means of a (Dektak 3030) Profilometer. Morphologic characteristics of the coatings like grain size and roughness were obtained by using atomic force microscopy (AFM) under contact mode from Asylum

Research MFP-3D® and calculated by a Scanning Probe Image Processor (SPIP®) In this work, the SPIP® was used in the grain size analysis for a quantitative studyof grains and particles, and in the roughness analysis used for an advanced measurement of the surface roughness. Surface characteristics were determined via scanning electron microscopy (SEM, Phenom FEI) equipped with an optic light with a magnification range of 525–24,000X and a high-sensibility detector (multi-mode) for scattering electrons. To determine chemical composition, an exhaustive XPS study was carried out for Hafnium with Metal transition Multilayers; thus, XPS was used to analyze the bonding of hafnium, vanadium, and nitrogen atoms by using ESCAPHI 5500 monochromatic Al-Kα radiation and a passing energy of 0.1 eV. Surface sensitivity of this technique is so high that any contamination can produce deviations from the real chemical composition and, therefore, the XPS analysis is typically carried outunder ultra-high vacuum conditions with a sputter cleaning source to remove any undesired contaminants.

Materials and Methods for Wear Tests Mechanical Engineering of Hafnium with Metal Transition Multilayers

Mechanical analyses were performed via nanoindentations by using an Ubi1-Hysitron device and a diamond Berkovich tip at variable loads. These results were evaluated through the Oliver and Pharrmethod. Tribological characterization was performed by means of Microtest, MT 400-98 tribometer. Adherence of the layers was studied by using a Scratch Test Microtest MTR2 system; the parameters were a 6-mm scratch length and a raising load of 0-90 N. Scanning electron microscopy was used to identify the different adherence failures.

RESULTS AND DISCUSSION

X-Ray Analyses of Hafnium with Metal Transition Multilayers

The total measured thickness of the deposited HfN/VN multilayered coating was approximately 1.2 μm in all cases. Individual thicknesses varied in function to the bilayer number from n = 1 to 80, producing layers with thicknesses from 1200 to 15 nm, respectively.

Figure 1 shows the high-angle X-ray diffraction patterns corresponding to the HfN/VN superlattices. Clear evolution of diffraction patterns is noted in this set of multilayers as bilayer period is increased. As seen in Figure 1, at large bilayer periods, in this investigation was found a close structural coherence between preferred orientation for Hf-N layer and FCC (200) preferred orientation for V–N layers (quasi-isostructural superlattices).

These preferential orientations agree with JCPDS00 033 0592 (HfN) and JCPDS00 035 0768 (VN) from ICCD cards. The texture of HfN layer remains constant in preferential orientation (111) from 1200-nmthick films to multilayers with 15-nmthickness; this behavior suggests the possibility of a quasi cube-on-cube epitaxial growth. In the thinnest bilayer pattern ($\Lambda \leq 15$ nm) there are still some small contributions of Hf–N (311) and (222) reflections, but they disappear for a large range of bilayer periods till, the thickest period multilayer ($\Lambda \leq 1200$ nm), where a great Hf–N (111) and small V-N (200) peak appears. On the other hand, in this research was found a little shift of diffraction patterns towards high angles (Figure 1c). So, if the shift of diffraction patterns is truly present in HfN/VN multilayers can be in relation to the compressive residual stress characteristic for those multilayered systems. Therefore, it was observed that the HfN (111) peak position suffers deviation from the bulk value, indicating possible stress evolution of HfN/VN layers with the bilayer period, moreover the stress into the multilayer can be associate to lattice mismatch (8.11%).

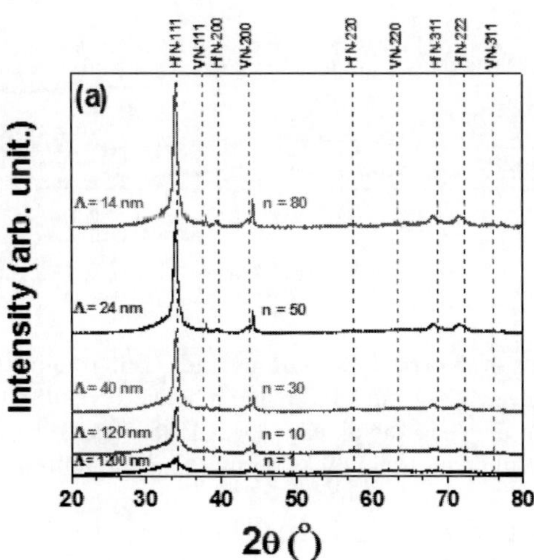

Figure 1. Continued on next page.

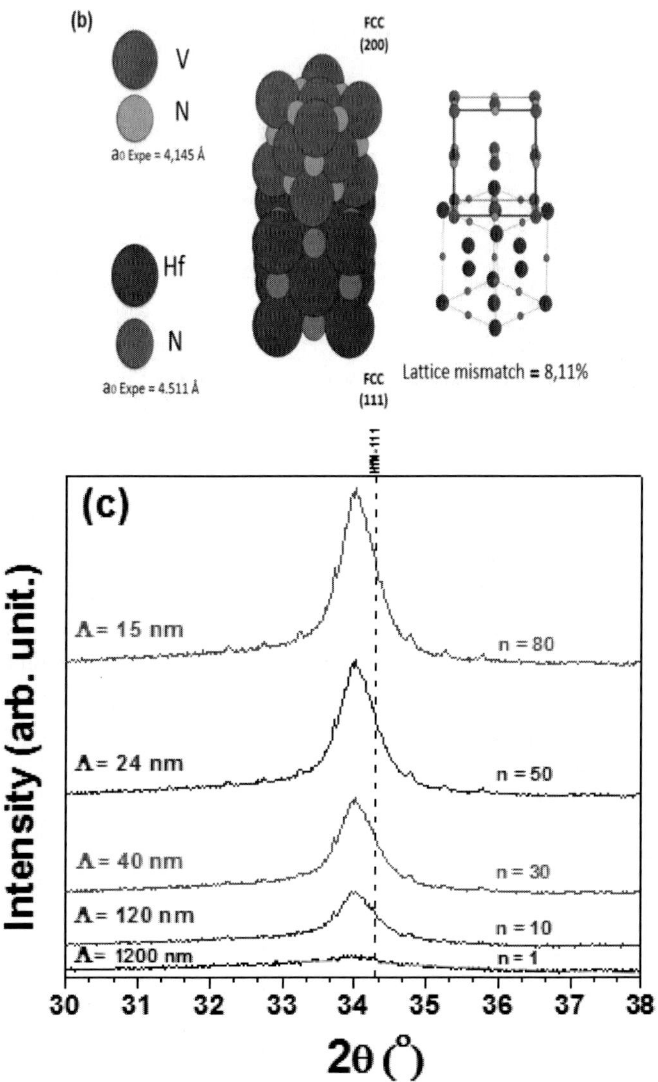

Figure 1. XRD patterns of the HfN/VN multilayered coatings deposited on Si (100) substrates with Λ between 1200 and 15 nm; and n between 1 and 80. Dash lines indicate the position of the peaks obtained from JCPDS files, (b) the (c) maximum peak with shift toward high angles in relationship to increasing bilayer number (n).

For thinner multilayered periods (n = 80, Λ= 15 nm), a continuous transition of Hf–N (111) peak position was observed, from multilayered films with bilayer periods of 120 nm to those with 15 nm bilayer period in agreement with crystal simulation for cube-on-cube assembly taking in account the HfN/VN mismatch (Figure 1c). (Figure 2c) show the quasi-relaxed position observed for thinner bilayer periods was progressively shifted to higher compressive stress values as the bilayer period increased until the $\Lambda = 1200$ nm value.

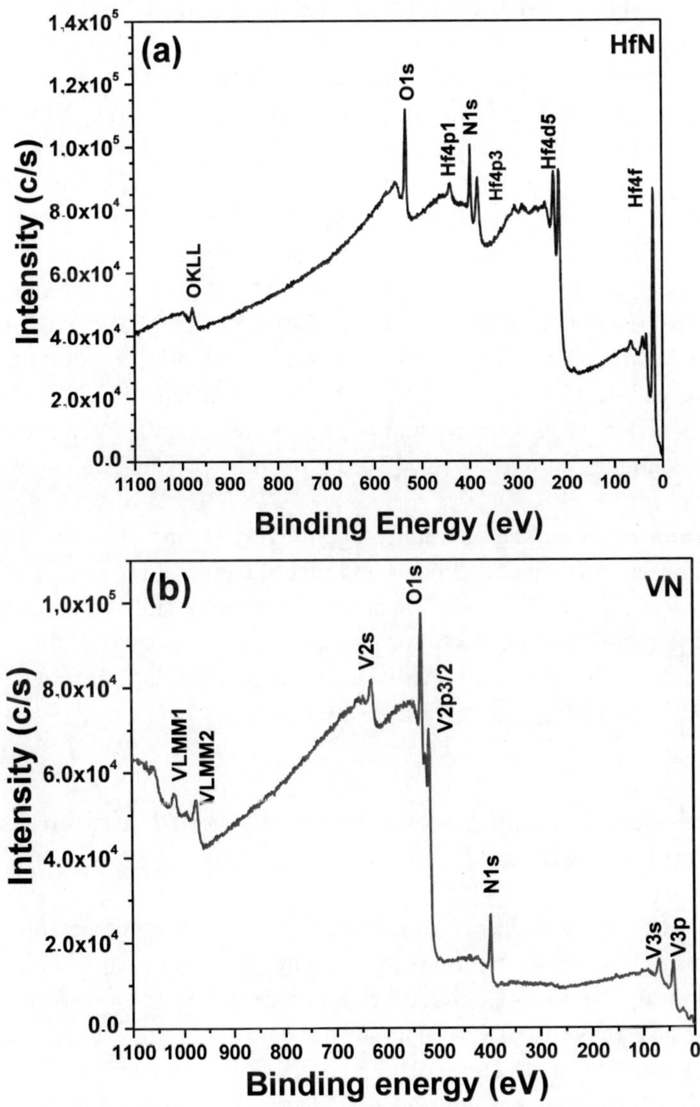

Figure 2. XPS survey spectrum (a) HfN coatings and (b) VN coatings deposited on Si with an r.f. negative biasvoltage of −30 V.

These effects presenting a stress relief due to the movement of this peak towards higher angles compared to other multilayers, but close to bulk value (34.01°).

XPS Analysis of Hafnium with Metal Transition Multilayers

The XPS survey spectra for HfN and VN single layers that make up the HfN/VN multilayered coatings are shown in Figure 2. For HfN material (Figure 2a) the peaks at 523.2 eV, 397.6 eV, 224.8 eV, and 18.4 eV correspond to O1s, N1s, Hf4d5, and Hf4f binding energies, respectively. The change of binding energy compared to HfN materials verifies the formation of Hf-N layer; therefore, calculating the peak area yields an atomic ratio of Hf:N =1.1:0.9, similar to the stoichiometry of $Hf_{1.1}N_{0.9}$ [10]. On the other hand, for VN material (Figure 2b) the peaks at 630.4 eV, 532.0 eV, 516.8 eV, and 397.6 eV correspond to V2s, O1s, V2p3/2, and N1s, binding energies, respectively. The change of binding energy compared to VN verifies the formation of binary V - N compound; therefore, calculating the peak area yields an atomic ratio of V:N = 1.1:0.9, similar to the stoichiometry of $V_{1.2}N_{0.8}$ [11].

According to the XPS literature regarding Hf-N and V-N materials [10, 11], the concentration measurements and identification of the specific bonding configurations for the HfN and VN layers are more reliable when is used the XPS analysis. So, the core electronic spectra carry information of the chemical composition and bonding characteristics of the HfN and VN films generate an increase in the reliability of the results.

Transmission Electron Microscopy Analysis of Hafnium with Metal Transition Multilayers

HfN/VN superlattices modulations and microstructures were accomplished by TEM micrographs. Figure 3 presents the TEM cross-sectional image of an HfN/VN superlattices with $\Lambda = 15$ nm (80 bilayered). The darkest contrast of HfN layers with respect to those of VN allowed clear determination of layer structures. These HfN/VN superlattices as hard coatings presented well-defined and uniform periodicity because the bilayer period in multilayers was confirmed by using TEM. All the multilayer stacks were resolved by TEM and confirmed quite precisely by the previously designed nominal values of bilayer thickness, as well as the total thickness. Transmission electron microscopy images show that VN layers are marginally thicker than those of HfN; they also confirmed that for each multilayer there is a different deviation of 0.3 layer thickness ratio.

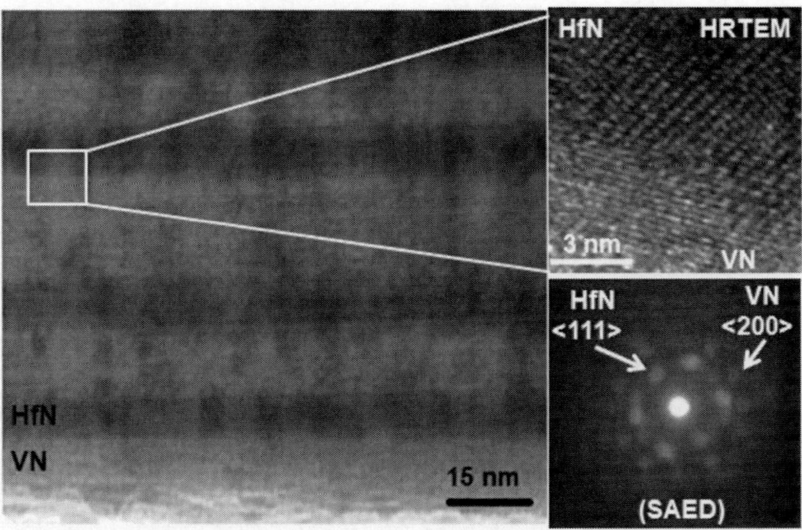

Figure 3. HfN/VN multilayer modulations by TEM image of coatings with n = 80, Λ = 15 nm with SAED pattern.

These multilayer modulations were possible and observed when the periodicity on the HfN/VN superlattices coating remained constant for the low bilayer periods (from Λ = 1200 to 15 nm), as shown via TEM results. Moreover, the TEM bright field micrograph in Figure 3 shows a multilayer with Λ = 14 nm, typical of thick bilayer periods. It reveals a compact crystalline structure with wide columnar grains extended along the entire multilayer stack.

The selected area electron diffraction pattern (SAED) of the whole multilayer indicated a (111)-preferred orientation for an HfN layer and (200)-preferred orientation for VN layer, confirming the structural phenomenon observed by XRD results. Also, the region shown in Figure 3 permitted studying the structure of a single crystallite, which can be identified as a small dark zone propagating throughout the multilayer. A SAED pattern of a region overlapping several periods of this structure shows an array of regularly distributed spots (Figure 3), confirming together with XRD results the expectations of a multilayer structure in (quasi-isostructural superlattices).

Atomic Force Microscopy Analysis of Hafnium with Metal Transition Multilayers

Atomic force microscopy was carried out to quantitatively study the surface morphology of our samples in relation to decreased bilayer periods or increased bilayer numbers in HfN/VN superlattices deposited onto Si (100). Figure 4 shows AFM images for multilayered coatings with statistical distribution of grain size that was analyzed by 1 μm×1 μm AFM images for (a) n = 1, (b) n = 10, (c) n = 30, (d) n = 50, and (e) n = 80. The correlation between grain size, part (a), and roughness, part (b), with the bilayer number is shown in Figs. 5a and b, respectively.

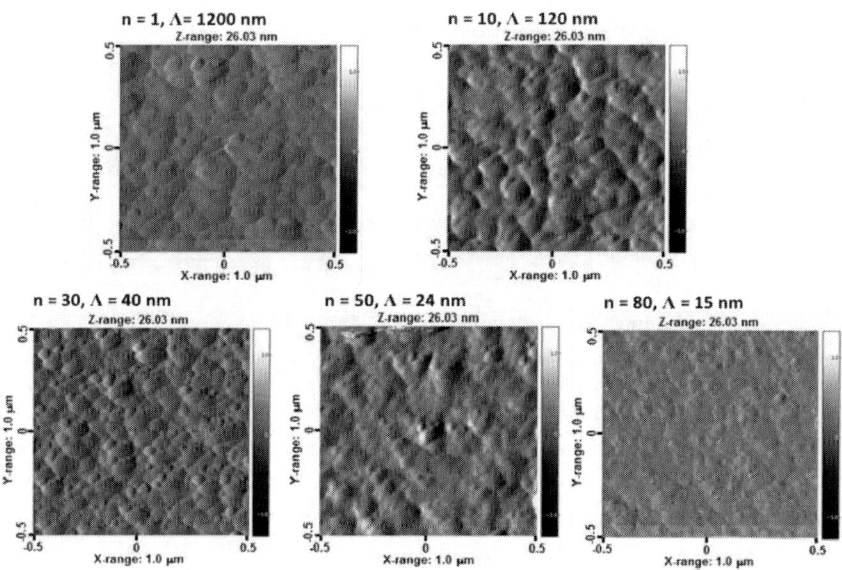

Figure 4. AFM images for an HfN/VN superlattices with n = 1, 10, 30, 50, and 80, (a-f) grown at r.f. negative bias voltage of −30 V.

The quantitative values were extracted from the AFM images by means of statistical analysis scanning probe image processor (SPIP®). The HfN/VN superlattices has the same total thickness, around 1.2 μm. Roughness was greater than for a coating with n = 1 or high bilayer period (Λ =1200 nm) and lower for multilayered coatings with a higher bilayer number (n = 80) or lower bilayer period (Λ = 15 nm), indicating that coatings with (n = 1) grow more disordered than do multilayered coatings with (n = 80). Grains of multilayered coatings are smaller than those for coatings with low bilayer period because the Ar+ ion bombardment on coatings stimulates a greater number of nucleation places, given the reduction of individual thicknesses for each layer when the bilayer period is reduced and bilayer numbers are increased; this implicates a decrease in the entire surface roughness [12, 13].

On the hands, it was possible take into account that grain size is affected by changes in grain size of inner layers HfN/VN as a function of reduced bilayer period (Λ) or increased bilayer number (n). The reduction of the spatial frequency (changes in the bilayer periods) may induce reduction in the residual stress (strain reliving) which can be observed (Figure 1c). Then, the changes in the bilayer periods, restrict growth phase modifying the superficial morphology (roughness) observed by AFM results and can also generate microstructural changes inducing greater topographic order.

Many authors have reported a correlation between AFM and XRD results because AFM analysis delivers surface information (with lateral resolution), XRD collects information from the space around the normal vector on the surface (from a cross-section); if this is taking into account, it is possible to consider that PVD coating columns can grow "vertically" from the substrate to the "top" and can change their cross-section when bilayer thickness is modified; therefore, reduced grain size and changes in surface roughness are evident.

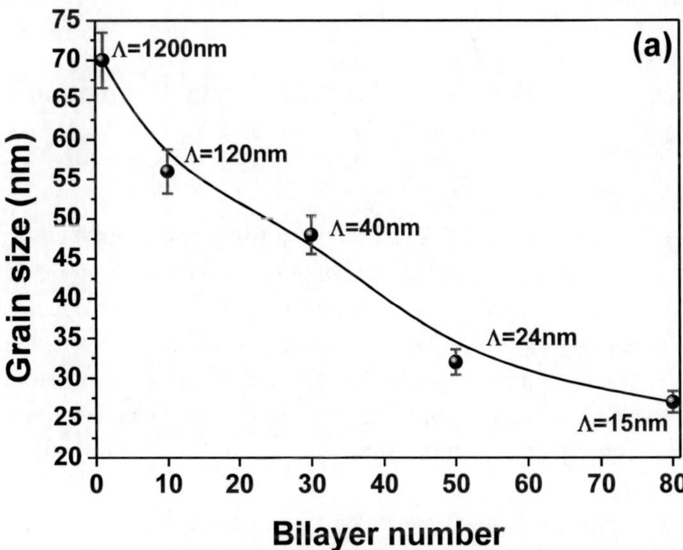

Figure 5. Continued on next page.

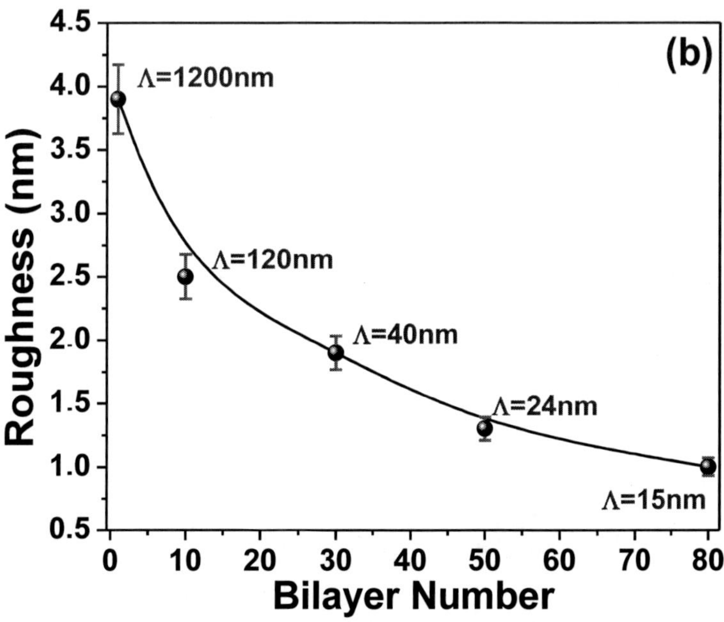

Figure 5. Surface measurements obtained via AFM and SPIP analysis for HfN/VN multilayers with a total thickness of 1.2 μm and bilayer period from Λ = 1200 to 15 nm: (a) correlation between grain size and bilayer number, and (b) correlation between roughness and bilayer number.

Mechanical Properties of Hafnium with Metal Transition Multilayers

Hardness and Elastic Modulus

Literature reports that the thickness period for many multilayered systems is between 5 and 20nm, which is relevant for mechanical properties [14]; however, in industrial applications, as mentioned in the introduction, it is necessary to obtain coatings with thicknesses above~1.0 μm, therefore, multilayer coatings deposited with bilayer period under 20 nm for a total thickness around ~1.2 μm imply a complex and extended number of multilayers; consequently, this study carry out a minimum bilayer period of 15 nm.

Thus, the typical load-displacement indentation curves of multilayer coatings was conducted by using the standard Berkovich indenter and indentation matrix image by AFM, as shown in Figure 6a and b. Values of elasticity modulus (Er) and hardness (H) were obtained by using Oliver and Pharr's method [15].

The experimental results associated with the mechanical properties correspond to the mixed hardness of the coated systems as a function of the relative indentation depth. As mentioned before, to compare the hardness response of the coatings, the work of indentation models described by Korsunsky et al. [16] was used to analyze the nanoindentation data. This model uses the following equation to describe the composite hardness and is based on the total work done by the indentation load to produce an indentation of depth (h):

$$H_C = H_S + \frac{H_f - H_S}{1 + K\beta^2}$$

(1)

In this model Hc is the apparent composite hardness, Hf is the intrinsic film hardness, Hs is the substrate hardness, k is a dimensionless materials parameter related to the composite response mode to indentation and $\beta = h/t$, which denotes the relative indentation depth with respect to the coating thickness, where h is the maximum indentation depth and t is the coating thickness.

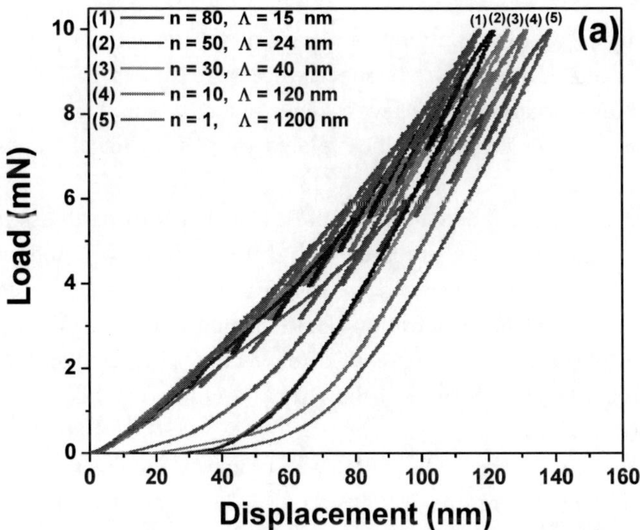

Figure 6. Continued on next page.

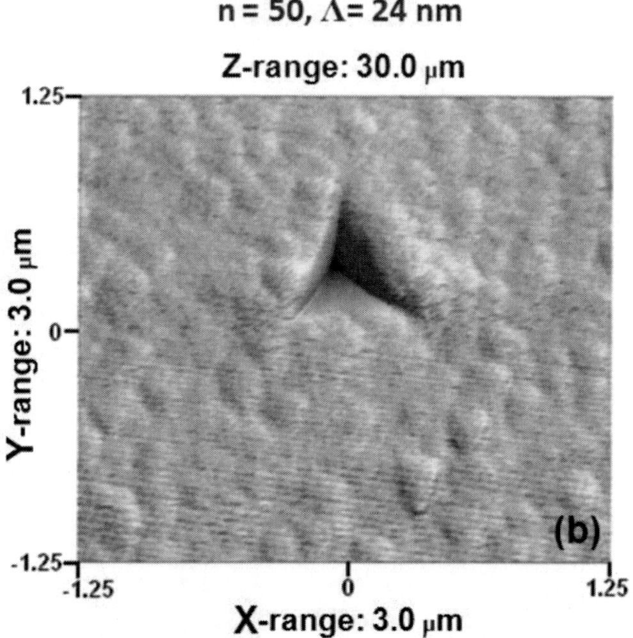

Figure 6. Nanoindentation results with (a) load-displacement indentation curves for all coatings and (b) indentation track images via AFM for multilayers with Λ= 24nm and n = 50.

The fitting results by applying the work of the indentation model to the experimental data of the HfN/VN superlattices are shown in Figure 7. The reduced elastic modulus here was determined from nano indentation measurements, considering the values related to 10% of the thickness of the coatings.

Normalized relative indentation depth (β = h/t) against hardness measured for the HfN/VN superlattices, prepared with different bilayer periods, is shown in Figure 7. All curves presented in Figure 7 show decreased composite hardness when increasing penetration depth, which is not surprising because the hardness of AISI 4140 steel is in all cases lower (~5 GPa) than those of the coatings. The hardness measured of the HfN/VN with n = 80 is clearly higher than that of HfN/VN superlattices with n = 1, 10, 30, 50, and 80. The greatest hardness of the multilayer set, 37GPa, was obtained for the thinnest bilayer period (Λ = 15 nm); therefore, it was 48and 43% greater than the values for the HfN and VN single-layer coatings, respectively. This effect is probable and attributed to many interfaces that blocked and deflected the micro-crack movements across the interface between HfN andVN layers due to differences in the shear module of the individual layer materials, as well as by coherency strains causing periodical strain-stress fields for lattice-mismatched multilayered coatings [16]. According to Gwang et al. [18] another condition toenhance the hardness of multilayered coatings with respect to the single layer is for the layer in multilayered systems to be discrete. This was shown in this study from TEM analyses (Figure 3). Enhancement of mechanical properties for HfN/VN superlattices results from increased hardness, as shown in multilayer-type coatings for diverse material systems (e.g., TiN/VN, TiCN/TiNbCN, and TiC/TiB2) [7, 13, 19]

The H values of the HfN/VN superlattices measured by nanoindentation results (Figure 6-7) are presented in Figure 8a, as functions of Λ and n. The Er values of the multilayered coatings are also presented in Figure 7, showing relevant differences in their values. The H and Er in these multilayered coatings varied from 25 to 37 GPa and from 265 to 352 GPa, respectively. The highest hardness of the multilayer set, 37 GPa, was obtained by the thinnest bilayer period ($\Lambda = 15$ nm) and it was 76and 94% greater than the expected reference value from the rule-of-mixtures applied to VN and HfN single-layer films, respectively.

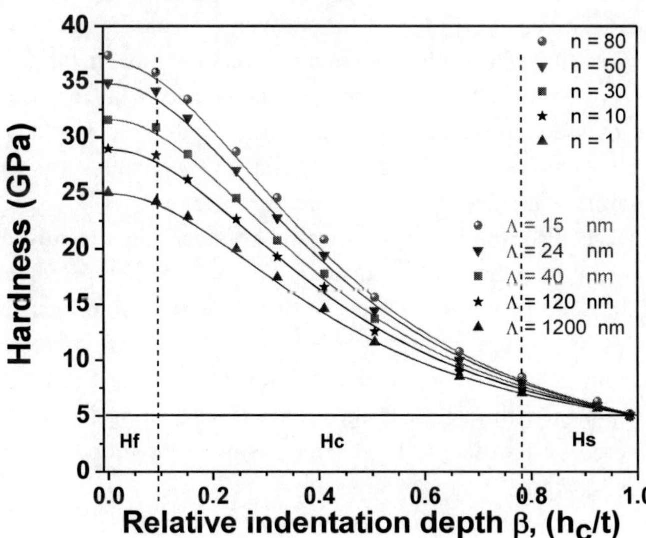

Figure 7. Relative indentation depths against the hardness measured for HfN/VN multilayered coatings, as a function of the bilayer number (n). Continuous lines correspond to the fitting of expression (1).

This increase in mechanical properties is related to the remarked superlattices effect present when a perfect assembly occurs for HfN and VN coatings deposited over an industrial AISI 4140 steel substrate. Enhancement of mechanical properties can be associated to improved hardness by using HfN/VN superlattices materials.

Many authors have used the Hall–Petch effect to explain the material hardening; therefore, it is possible to apply the Hall–Petch effect when the ceramic materials as hard multilayered coatings with $\Lambda > 5.2$ nm are obtained because multilayered coatings with lower Λ are within the nanoscale regime and, thus, the dislocations should not occur in nanoscale structures below certain grain size value [20].

Hardness for the HfN/VN superlattices is in relation to the reduction of the modulation period (Figure 8a). Therefore, from nanoindentation results (Figure 6), it is possible to observe that indentation test exhibit low deformation size when mechanical properties are enhance as function of increased bilayer number.

It is argued that there are some correlations between the hardness of multilayered coatings and the modulation period [14]. Hence, the hardness enhancement of HfN/VN is related to the evidence of a stress relieving according to interfaces distance decreased, as shown from XRD results (Figure 1) and TEM results (Figure 3); moreover, this mechanical effect was attributed to many interfaces blocking the dislocation movement across the interface between the HfN layer and the VN layer due to differences in the shear module of the individual layer materials, and by coherency strain, causing periodical strain-stress fields $(-\sigma HfN + \sigma VN)$ in the case of lattice mismatched multilayered coatings.

Each interface also functions as a grain boundary in a Hall-Petch [2] related mechanism like that of the dislocation pile-up that offers a strong interaction again with interfaces in general; therefore, each interface serves as crack tip deflector, which improves coating mechanical properties [21]. According to Kim et al., [22] another condition to enhance the hardness of multilayered coatings with respect to the single-layer coating is that the layer in the multilayered system has to be discrete, which was found in this study via TEM analyses (Figure 2).

Also, enhancement of mechanical properties for an HfN/VN multilayer is according to increased hardness, as shown in multilayered-type coatings for diverse material systems (e.g., TiN/VN, TiCN/TiNbCN, and TiC/TiB2) [7, 13, 19].

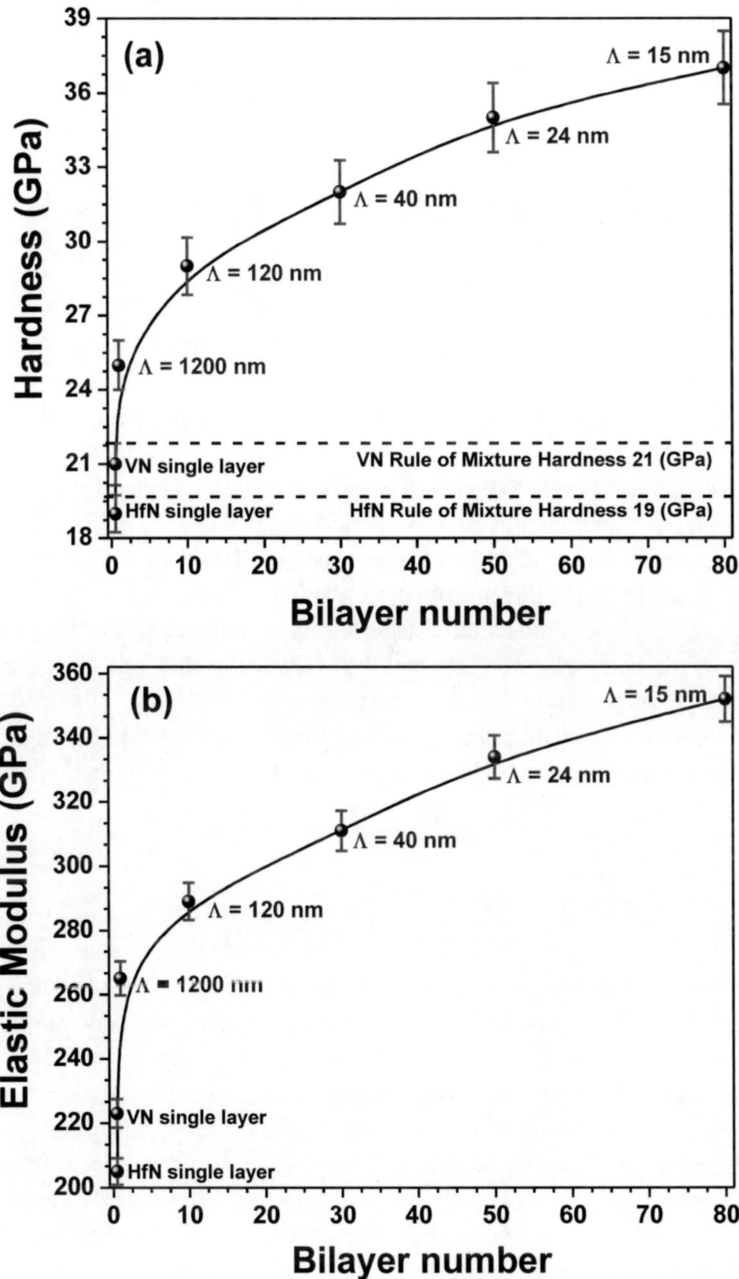

Figure 8. Mechanical properties for HfN/VN superlattices with Λ = from 1200 to 15 nm: (a) hardness as a function of n. and (b) elastic modulusas a function of n.

Yashar et al. [2] used a Hall-Petch approach to model the mechanical behavior of multilayered materials with layer thicknesses as low as 1-100 nm, transforming the Hall-Petch approach into the following equation presented by Caicedo et al., [13]:

$$H_m = H_{(f1+f2)} + k_{(IM)}D_{(t)}^{-\frac{1}{2}}$$

(2)

where H_m is the multilayer hardness, $H_{(f1+f2)}$ is the hardness from layer 1 and layer 2, k_{IM} is a constant measuring the relative hardening contribution of the interface between layer 1 and layer 2, and Dt is the bilayer period (Λ).

The model predicts the overall behavior of the hardness on Λ observed in most multilayered systems. Previous work has shown that a maximum hardness would be expected when the individual components of the multilayer have a relative equal thickness, as presented in this work [23].

Furthermore, dependence of the elasticity behavior on n or Λ was observed; the highest bilayer number corresponds to the highest Er; therefore, it is possible to conclude that elasticity and elastic recovery (R %) of the multilayered coatings are enhance by increasing the interface number. Thereby, from the nanoindentation measurement, the typical values of Er and H were obtained by using the Oliver and Pharr method [15]. The R% for all HfN/VN superlattices was calculated by using the following equation:

$$R = \frac{\delta_{max} - \delta_p}{\delta_{max}}$$

(3)

where δ_{max} is the maximum displacement and δ_p is the residual or plastic displacement. The equation data were taken from the load-penetration depth curves of indentations for each coating according to Figure 6. Figure 9a shows the increase in the elastic recovery as a function of increased bilayer number when δ_{max} and δ_p values from load and displacement results (Figure 6) are introduced in Eq. (3); thus, it is possible to observe the interface effect on a multilayered system because when bilayer number (n) increases the interface number is increased; therefore, this can generate hardening on multilayered hard coatings, evidenced by improved elastic recovery (R%).

Also, according to Kim et al., [22] a relationship exists between Er and H, known as the plastic deformation resistance (H3/E2) ratio; this relation was calculated for all multilayered coatings in function of bilayer number or bilayer period. Figure 9b shows a considerable increase in the resistance to plastic deformation (H3/E2) ratio as a function of increased bilayer number (n), this fact is due to the hardness and the elasticity modulus also increasing as the interface number increased for all multilayered coatings.

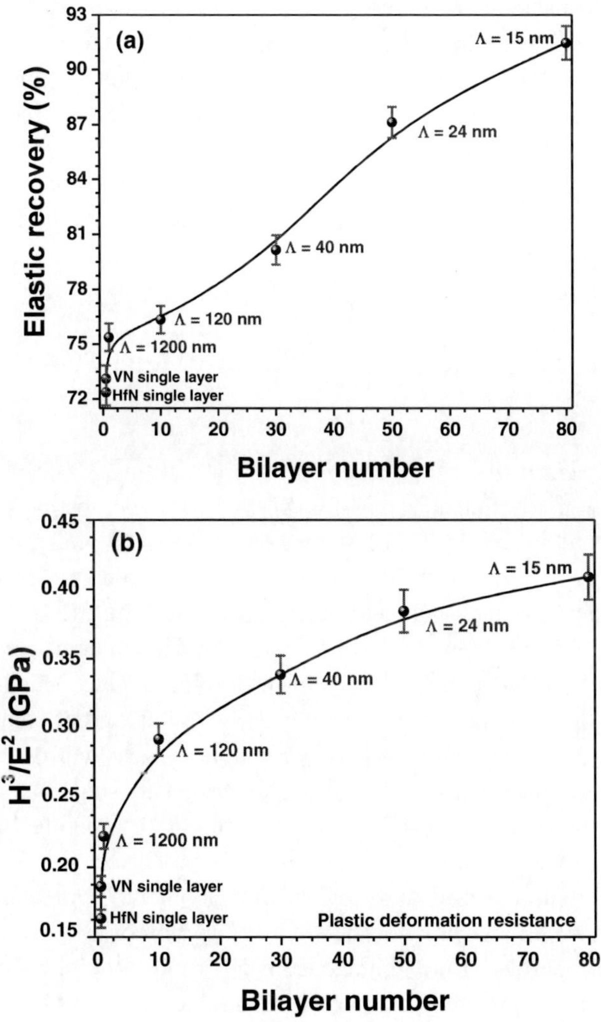

Figure 9. Elasto-plastic properties with plastic deformation for HfN/VN superlattices with bilayer period from Λ = 1200 to 15 nm: (a) R%, and (b) plastic deformation resistance.

This enhancement in plastic deformation resistance occurs when Λ decreases, increasing the interface number for coatings with total thickness; thus, producing the crystallite refinement, point defect formations, and increased interface number, which improves hardness of HfN/VN superlattices.

As shown in Figure 9, the HfN/VN superlattices increased plastic deformation resistance and elastic recovery with respect to coatings deposited with lower bilayer numbers. The maximum value was reached for n = 80 and Λ= 15 nm, i.e., plastic deformation due to applied load is more markedly reduced than that of other multilayered systems with fewer bilayer numbers. This effect clearly correlates to increasing coating density, hardness, and elastic recovery [15]. In general, the bilayer number (n) or bilayer period (Λ) has some effects on the multilayer shown to exhibit very high hardness in nitride multilayered coatings.

Tribological Properties of Hafnium with Metal Transition Multilayers

Pin-on-Disk Analysis

The friction coefficient values for AISI 4140 steel substrates coated with all multilayered systems (Λ = 1200 mm, n= 1; Λ = 120 nm, n = 10;Λ = 150 nm, n = 30; Λ = 40 nm, n = 50; Λ = 24 nm, n = 50; and Λ = 15 nm, n = 80) were tested against steel balls and are presented in Figure 10a. These curves showed two distinct stages. In the first stage, the friction coefficient (μ) began at a low level (0.15–0.25) (Figure 10a) in the first contact; this stage can be attributcd to the running-in period associated with the kind of contact between the steel ball and the coating, where formation of wear debris occurs by the cracking of roughness tips on both counterparts. This stage has a short time period, then the friction coefficient increases to 0.4–0.5 (Figure 10a) followed by a decrease to the friction coefficient of stage II. This stage is defined as the steady-state friction period and begins after about 30–80m of sliding [19]. Figure 10b shows the friction coefficient as a function of bilayer numbers. The tribological properties of the homogeneous HfN and VN single layer coatings were provided in Figure 10 for comparison in relation to multilayered systems. In these tribological results showed the reduction of the friction coefficient while the bilayer number increased and the bilayer period decreased. The friction coefficient of HfN/VN superlattices ranged from approximately 0.33 to 0.15, being the lowest value reported for the multilayer growth with Λ = 15 nm, n = 80. The friction coefficient value represented a decrease at approximately 75and 65% of the friction coefficient with respect to the HfN and VN single layers, respectively. The last behavior can be related to the friction mechanical model proposed by Archard [24], which relates the contribution of the contact surface roughness and the elastic–plastic properties of the coating in the following equation:

$$\mu = \frac{F_f}{F_n} = C_k \cdot \frac{R_{(s,a)}}{\sigma t_{(H,Er)}}$$

(4)

where μ is the friction coefficient, Ck is the constant that depends on the parameter of the test, R(s,a) is coating roughness, and σt is a variable that takes into account the elastic–plastic properties (hardness, H, or elastic modulus, Er), obtained by mechanical measures [24]. In agreement with the Archard model, when the surface coating has low roughness and high hardness, the friction coefficient will tend to decrease and will be stable for long sliding distances, specifically if the counterpart of the test is softer than the coating. On the other hand, although hardness has long been regarded as a primary material property that defines wear resistance, strong evidence suggests that the elastic modulus can also have an important influence on wear behavior. In particular, the elastic strain to failure, related to the ratio of hardness (H) and elastic modulus (Er), which has been shown by a number of authors to be a more suitable parameter to predict wear resistance than with hardness alone. Until now, scientific research has been mainly aimed at achieving ultra-high hardness associated with high elastic modulus, the latter of which, conventional fracture mechanics theory would suggest, is also desirable for wear improvement (by preventing crack propagation). This study discusses the concept of multilayered coatings with relatively high hardness and high elastic modulus, which can exhibit improved toughness and are, therefore, better suited to optimize the wear resistance of 'real' industrial substrate materials (i.e., steels and light alloys with low moduli). Recent advances in the development of ceramic–ceramic HfN/VN superlattices are summarized and discussed in terms of their relevance to practical applications. This paper observed that the elastic strain to failure (which is related to H3/E2) affect the tribological behavior in HfN/VN superlattices that, although not necessarily exhibiting extreme hardness, provide superior wear resistance when deposited on the types of substrate materials which industry needs to use [25].

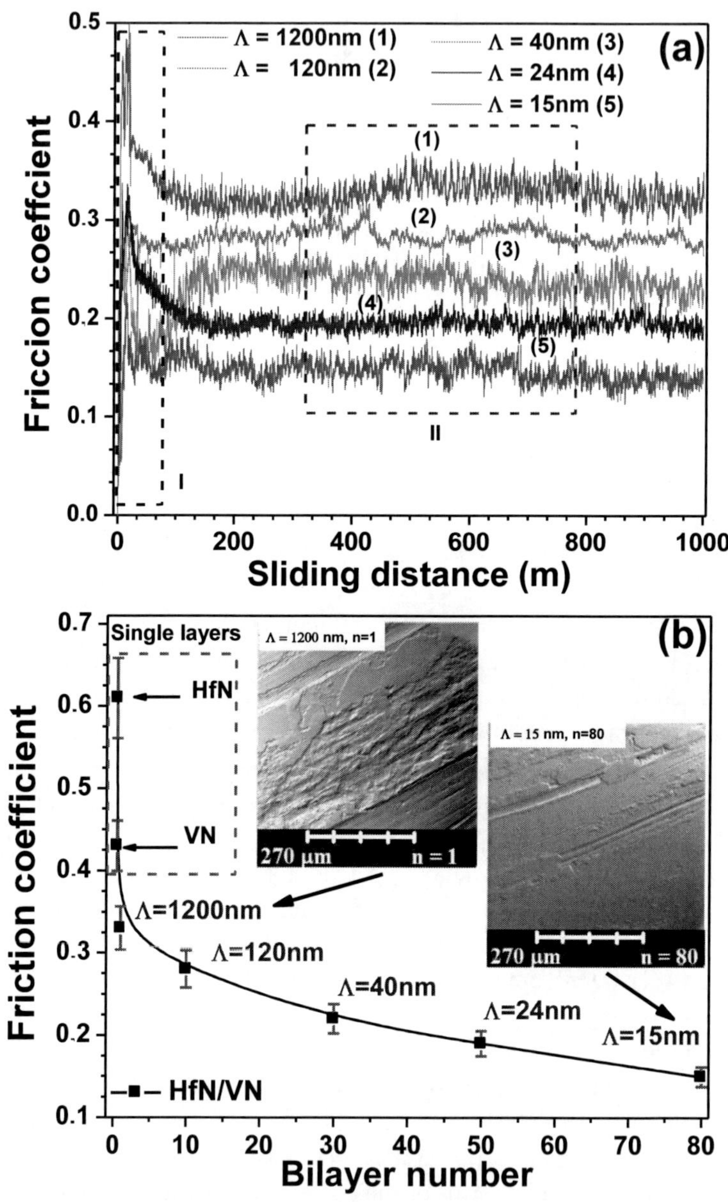

Figure 10. Tribological results of 4140 steel substrates coated with HfN/VN superlattices: (a) friction coefficient as a function of sliding distance and (b) friction coefficient as a function of the bilayer numbers (n) or bilayer periods (Λ) with wear track observed by SEM.

Therefore, this behavior suggests that improving plastic deformation resistance (H3/E2) when the bilayer number is increased exert more wear resistance due to enhanced mechanical properties (Figure 9) associated to perfect modulation assembly observed from XRD (Figure 1) and TEM results (Figure 3); thus, generating a reduction in the friction coefficient (Figure 10).

In the insert of Figure 10b from SEM results it is possible to find different wear mechanisms, such as: abrasion, adhesion, oxidation, and diffusion. In this sense, the abrasion mechanism is a predominant phenomenon at multilayers with low bilayer number (low mechanical properties). The values for multilayer period at which the maximum wear values occur will depend on different factors like the combination of high roughness, small grain size, and low elastic modulus, among others.

ADHESION BEHAVIOR OF HAFNIUM WITH METAL TRANSITION MULTILAYERS

The scratch test technique was used to characterize coating adherence strength. The adhesion properties of single-layer coatings and multilayered coatings can be characterized by the following two terms: LC1, the lower critical load, which is defined as the load where cracks first occurred (cohesive failure) and the LC2, the upper critical load, which is the load where the first delaminating at the edge of the scratch track occurred (adhesive failure) [23]. The values of critical load (LC1 and LC2) for the different coatings are shown in Figure 11. The LC1 was shown for the different coatings in the range of 27–58 N; the lowest value was attributed to the multilayered coating deposited with n = 1 and the highest value was attributed to the HfN/VN superlattices growth with $\Lambda = 15$ nm and n = 80.

The critical loads in adhesive failure (LC2) values for the different coatings are summarized in Figure 12. This figure clearly shows the increased adhesion properties of HfN/VN superlattices as a function of decreased bilayer period. Due to the quantitative adhesion measurements between layers and substrates, it is a complex process even for single-layer coatings in agreement with some authors [19]; a qualitative characterization is necessary to evaluate the adhesion behavior for all multilayered systems, as described before in this section, i.e., in terms of LC1 and LC2 critical loads. Therefore, for the purpose of ensuring a fair comparison between the different coating systems, it was assumed that the adhesion between the substrate and the first layer of the multilayered system remains constant because the preparation conditions of samples and the coating deposition parameters used were the same. Besides, in all cases, it was verified that the parameters of the scratch test for all samples were also the same. According to the latter, it was expected that the response to the applied load will only depend on the coating properties because of the effect of each layer and the interfaces that make up the entire multilayered system. From Figure 12, it was observed that critical load increased when bilayer period (Λ) is reduced and bilayer number (n) is increased; this improvement is in part due to increased coating/substrate deformation resistance. In this mechanism, each interface serves as crack tip deflectors that change the direction of the initial crack when it penetrates deep into the coating and strengthens the coating systems. Moreover, by decreasing the bilayer period the dislocations among the layers found a major impediment to moving; therefore, those will require higher critical shear stress to move and spread throughout the coating and allow delaminating the coating. This means that multilayered coatings fail in laminar manner [14] due to the multilayered assembly, as found in the TEM analysis (Figure 3).

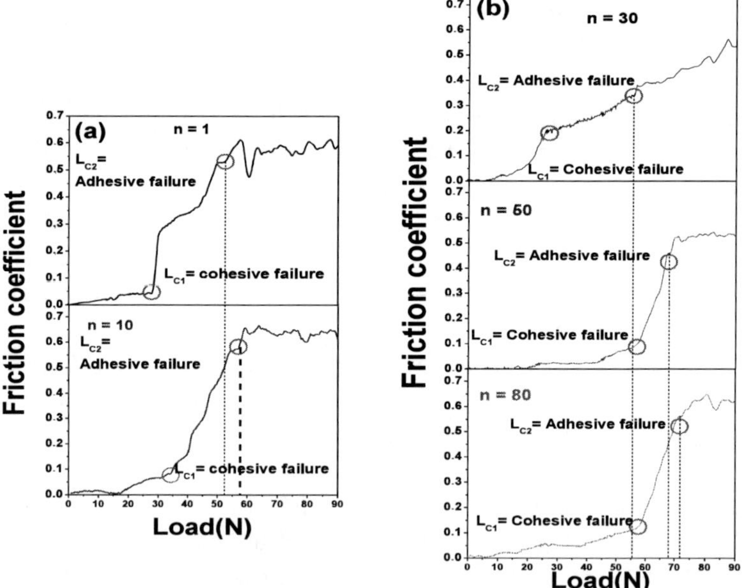

Figure 11. Tribological results for friction coefficient curves versus load for HfN/VN multilayered coatings, showing the adhesion failure (LC2): (a) Λ = 1200 nm, n = 1; Λ = 120 nm, n = 10; (b) Λ = 40 nm, n = 30; Λ = 24 nm, n = 50; and Λ = 15 nm, n = 80.

In consequence, multilayered and multiple structures like those studied in this research can enhance the resistance of coatings against crack propagation in relation to mechanical property evolution presented by enhanced hardness and elastic modulus (Figure 8) with highest elastic recovery (%R) (Figure 9a), preserving the integrity of the coatings under punctual and dynamic loads [26]. In this work, it was observed that an increase of 27% in the LC2 for HfN/VN superlattices with $\Lambda = 15$ nm and n = 80 in relation to multilayers with lowest bilayer number (n = 1).

Surface Tribological Analysis of Hafnium with Metal Transition Multilayers

Scanning electron microscopy images showing the different behaviors of the multilayered coatings after scratch tests are shown in Figure 13a–b. these images, revealed that at the beginning of the scratch pronounced deformation appeared due to the substrate plastic deformation and a coating debris removal, associated to the adhesive layer/substrate failure mechanism sideward lateral flanking [27]; thus, the SEM images agree with the scratch test results observed in Figure 13.

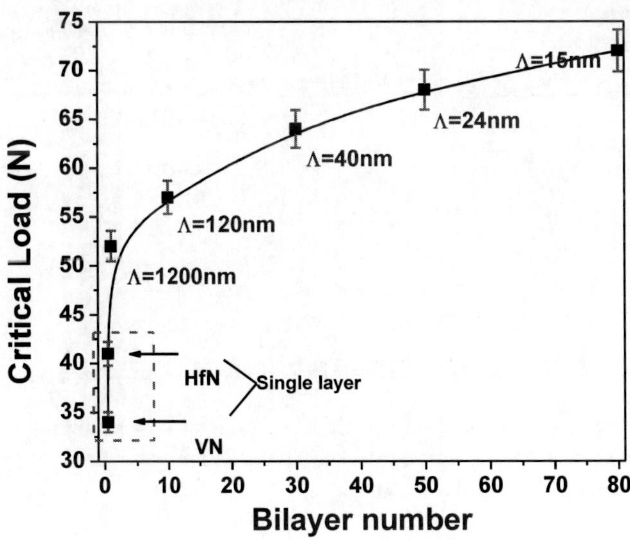

Figure 12. Correlation of critical loads (adhesive failure) with bilayer numbers (n) or bilayer periods (Λ).

Figure 13a shows a premature adhesion failure due to accumulation of stress at the scratch edges.to the multilayered system deposited with n = 1 (Λ = 1200 nm). Performing a detailed analysis of failure mechanisms for the HfN/VN superlattices with n = 80 (Λ = 15 nm) at the beginning of the scratch mark, there was a pronounced irregularity due to plastic deformation of the metal substrate (Figure 13e).

Later (furthest from scratch), conformal cracking of the layer associated with adhesive failure (LC2) appears. As other multilayered systems (n = 1, 10, 30 and 50), the systems with 80 bilayered coatings also presented "Recovery Spallation" type wear mechanisms (Figure 13e).

These series of multilayers showed a failure behavior type "buckling Cracks" that is characteristic of protective systems where the substrate is ductile and hard coating have good adhesion between them, these systems generate compressive efforts that are characteristic of cracks buckling failure mode [27] .

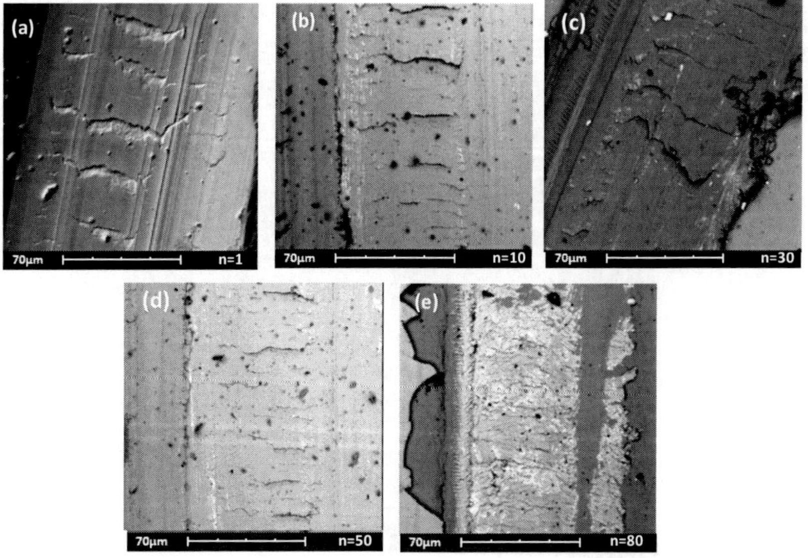

Figure 13. SEM micrographs of scratch tracks on (a) n =1(Λ = 1200 nm); (b) n = 10 (Λ = 120 nm); (c) n = 30 (Λ = 40 nm); (d) n = 50 (Λ = 24 nm); and (e) n = 80 (Λ = 15 nm) deposited onto industrial AISI 4140 steel.

CONCLUSION OF HAFNIUM WITH METAL TRANSITION MULTILAYERS

A nanometric HfN/VN superlattices structure was identified as a quasi-quasi-quasi-isostructural multilayer of HfN and VN phases. The preferential orientation for FCC HfN (111) and FCC (200) of VN single layer was shown via XRD. From XPS results, it was possible to identify the chemical composition in both single-layer coatings. The TEM analysis confirmed well-defined multilayered structures and showed slight variations in layer thickness ratio. The morphology and surface quality was determined for all coating systems, establishing homogeneous surfaces for all coatings. The grains of the multilayer were smaller than those of the individual HfN and VN coatings; this fact was attributed to the ion bombardment of the coatings, which stimulated a greater number of nucleation sites that together with grain size reduction by decreased bilayer period led to the entire surface roughness decrease.

It was found that the highest values of hardness and elastic modulus, 37 GPa and 351 GPa, respectively, were observed for multilayered systems with Λ = 15 nm and n = 80. Enhanced hardness of the HfN/VN stack was attributed to many interfaces blocking the micro-crack movements across interfaces between the HfN layer and VN layer due to differences in the shear module of the individual layer material, and by coherency strain causing periodical strain-stress fields, which is characteristic in those lattice-mismatched multilayered coatings, together with the Hall-Petch models, which give a good overall picture of the hardness enhancements observed.

High tribological performance with critical loads in adhesive failure of 72 N and friction coefficient of 0.18 were observed for multilayered systems with Λ = 15 nm and n = 80. From SEM micrographs, it was possible to determine that for the multilayered coatings different types of adhesive layer/substrate failures appear under strong plastic deformation conditions, which is important for the manufacture of wear resistant cutting and forming tools and mechanic devices used in industrial applications.

ACKNOWLEDGMENTS

This research was supported by "El patrimonio Autónomo Fondo Nacional de Financiamiento para la Ciencia, la Tecnología y la Innovación Francisco José de Caldas" under contract RC-No. 275-2011 and the program

"Jóvenes Investigadores e Innovadores Virginia Gutiérrez de Pineda No. 525-2011". Moreover, the authors acknowledge the Center of Excellence for Novel Materials (CENM), Universidad Militar Nueva Granada and, CINVESTAV, Mexico. Moreover, the authors acknowledge the Serveis Científico-Tècnics of the Universitat de Barcelona for XRD, XPS, and TEM analysis.

REFERENCES

[1]   Ghosh S. K., P. K. Limaye, B. P. Swain, N. L. Soni, R. G. Agrawal, R. O. Dusane, and a. K. Grover, "Tribological behaviour and residual stress of electrodeposited Ni/Cu multilayer films on stainless steel substrate," *Surface and Coatings Technology.*, 2007, 201, p 4609–4618.

[2]   Barshilia H. C., A. Jain, and K. S. Rajam, "Structure, hardness and thermal stability of nanolayeredTiN/CrN multilayer coatings." *Vacuum.*, 2003. 72, p 241–248.

[3]   Morant C., D. Cáceres, J. M. Sanz, and E. Elizalde, "Nano-mechanical properties of BCN/CN/BN multilayer films." *Diamond and Related Materials.*, 2007, 16, p 1441–1444.

[4]   Kim S., "Analysis of interfacial strengthening from composite hardness of TiN/VN and TiN/NbN multilayer hard coatings." *Surface and Coatings Technology,* 2004, 187 p 47–53.

[5]   Wu F.-B., S.-K.Tien, and J.-G. Duh, "Manufacture, microstructure and mechanical properties of CrWN and CrN/WN nanolayered coatings." *Surface and Coatings Technology,* 2005, 200 p 1514–1518.

[6]   Chan Y.-C., H.-W.Chen, P.-S.Chao, J.-G.Duh, and J.-W. Lee, "Microstructure control in TiAlN/SiNx multilayers with appropriate thickness ratios for improvement of hardness and anti-corrosion characteristics." *Vacuum.*, 2013, 87 p 195.

[7]   Yu L., S. Dong, J. Xu, and I. Kojima, "Microstructure and hardening mechanisms in a-Si3N4/nc-TiN nanostructured multilayers." *Thin Solid Films.*, 2008, 516 p 1864–1870.

[8]   Nordin M., M. Larsson, and S. Hogmark, "Mechanical and tribological properties of multilayered PVDTiN/CrN, TiN/MoN, TiN/NbN and TiN/TaN coatings on cemented carbide." *Surface and Coatings Technology.*, 1998, 106, p 234–241.

[9]   Tsai Y.-Z and J.-G. Duh, "Tribological behavior of CrN/WN multilayer coatings grown by ion-beam assisted deposition." *Surface and Coatings Technology.*, 2006, 201, p 4266–4272.

[10] Arranz A., "Synthesis of hafnium nitride films by 0.5–5 keV nitrogen implantation of metallic Hf: an X-ray photoelectron spectroscopy and factor analysis study." *Surface Science.*, 2004, 563 p 1–12.

[11] Glaser A., S. Surnev, M. G. Ramsey, P. Lazar, J. Redinger, R. Podloucky, and F. P. Netzer, "The growth of epitaxial VN(111) nanolayer surfaces." *Surface Science,* vol. 601, no. 21, pp. 4817–4823, Nov. 2007.

[12] Gómez M.E., J. Santamaria, S. Kim, M. Kannan, I.K. Krishnan, Schuller. "Detailed structural analysis of epitaxial MBE-grown Fe/Cr superlattices by x-ray diffraction and transmission-electron spectroscopy". *Phys. Rev. B.,* 2005, 71 p125410.

[13] Caicedo J. C., C. Amaya, L. Yate, M. E. Gómez, G. Zambrano, J. Alvarado-Rivera, J. Muñoz-Saldaña, and P. Prieto, "TiCN/TiNbCN multilayer coatings with enhanced mechanical properties," *Applied Surface Science.*, 2010, 256 p. 5898–5904.

[14] Stueber M., H. Holleck, H. Leiste, K. Seemann, S. Ulrich, and C. Ziebert, "Concepts for the design of advanced nanoscale PVD multilayer protective thin films." *Journal of Alloys and Compounds.*, 2009, 483 p 321–333.

[15] Oliver W. c. and G. m. Pharr, "An Improved Technique for Determining Hardness and Elastic Modulus Using Load and Displacement Sensing Indentation Experiments," *Journal of Materials Research.* 1992, 7 p 1564–1583.

[16] Korsunsky A. M., M. R. McGurk, S. J. Bull, and T. F. Page, "On the hardness of coated systems." *Coatings Technology.*, 1998, 99, p 171–183.

[17] Jehn H, "Improvement of the corrosion resistance of PVD hard coating–substrate systems." *Surface and Coatings Technology.*, 2000, 125 p. 212–217.

[18] Kim G. S., S. Y. Lee, and J. H. Hahn, "Synthesis of CrN/AlN superlattice coatings using closed-field unbalanced magnetron sputtering process," *Surface and Coatings Technology.*, 2002, 171 p. 91–95.

[19] Holleck H., M. Lahres, and P. Woll, "Multilayer coatings—influence of fabrication parameters on constitution and properties." *Surface and Coatings Technology.*, 1990, 41, p. 179–190.

[20] Park J.-K and Y.-J.Baik, "The crystalline structure, hardness and thermal stability of AlN/CrN superlattice coating prepared by D.C. magnetron sputtering." *Surface and Coatings Technology.*, 2005, 200 p. 1519–1523.

[21] Wang X. C., W. B. Mi, E. Y. Jiang, and H. L. Bai, "Structure and mechanical properties of titanium nitride/carbon nitride multilayers." *Applied Surface Science,* 2009, 255 p. 4005–4010.

[22] Kim G. S., S. Y. Lee, and J. H. Hahn, "Synthesis of CrN/AlN superlattice coatings using closed-field unbalanced magnetron sputtering process." *Surface and Coatings Technology.,* 2002, 171, p 91–95.

[23] Lewis D. B., I. Wadsworth, W. Mu, R. Kuzeljr, and V. Valvoda, "Structure and stress of TiAlN/CrN superlattice coatings as a function." *Surface and Coatings Technology.,* 1999, 119, p 284–291.

[24] Archard J. F., "Contact and Rubbing of Flat Surfaces," *Journal of Applied Physics.,* 1953, 24 p. 981.

[25] Leyland A and A. Matthews, "On the significance of the H/E ratio in wear control: a nanocomposite coating approach to optimized tribological behaviour," *Wear*, 2000, 246, p. 1–11.

[26] Podgornik B., S. Hogmark, O. Sandberg, and V. Leskovsek, "Wear resistance and anti-sticking properties of duplex treated forming tool steel." *Wear.,* 2003, 254p. 1113–1121.

[27] Holmberg K., H. Ronkainen, and A. Matthews, "Tribology of thin coatings," *Ceramics International*, 2000, 26, p. 787–795.

In: Hafnium
Editor: HongYu Yu

ISBN: 978-1-63463-164-8
© 2015 Nova Science Publishers, Inc.

Chapter 3

HAFNIUM CARBIDE COATING: PROPERTIES OF BULK, SURFACE AND METAL/HfC INTERFACES

H. Si Abdelkader and H. I. Faraoun*

Division Etude et Prédiction des Matériaux,
Unité de Recherche Matériaux et Energies Renouvelables.
LEPM-URMER. Université de Tlemcen – Algérie

ABSTRACT

This chapter considers basic research related to the Hafnium carbide coatings. We present results of physical properties obtained from a number of density functional theory studies, including our own. We firstly expose the bulk properties; the structural and electronic properties of HfC, Hf_3C_2 and Hf_6C_5 are illustrated and discussed. Elastic properties are also reported in order to assess the mechanical stability. Then we concentrate on the HfC surface properties. Where the surface energies and electronic properties of (100) orientation are calculated.

Finally, we discuss the atomic structure, bonding, adhesion and magnetism of the Metal/HfC interface.

Keywords: Hafnium carbide, Surface and interface, Electronic structure, Ab initio calculations

---

* Corresponding author address: Email: hayet.siabdelkader@mail.univ-tlemcen.dz

INTRODUCTION

Transition Metals (TMs) of group VI are chemically very similar. They occur naturally in ores. However, hafnium (Hf) has a high thermal neutron absorption cross section in contrast to the very low one of zirconium, excellent corrosion resistance to many media [1]. Interpenetration of the transition metals of group VI and carbon characterize the group IV carbides, known as ultra-high temperature ceramics (UHTCs), are candidates for applications in the atmosphere of extreme thermal and chemical environments [2-5].

Transition metal carbides (TMCs) are metallic compounds with unique physical and chemical properties including an extremely high melting point and hardness, chemical stability, corrosion resistance combined with metallic electrical and thermal conductivities [3-8]. These features give transition metal carbides the capability to withstand high temperatures in oxidizing environments, making them candidates for applications in the atmosphere of extreme thermal and chemical environments [3,4]. Due to these interesting properties, they are widely used in industrial applications such as cutting tools, dental drills; rock drills in mining, abrasives, and hard coatings [9], optical coatings, electrical contacts, and diffusion barriers such as Hafnium carbide [10].

Hafnium carbide (HfC) is one of the most optimal protective coating materials due to the highest melting point, high hardness, good resistance to oxidation and ablation, low thermal conductivity, low diffusion coefficient of oxygen and low surface vapor pressure [11-13]. Hafnium carbide used in the production of whiskers (with nickel catalyst), coating for superalloys, coating on cemented carbide, and oxidation resistant coatings for carbon-carbon composites (co-deposited with SiC). In addition, HfC could be used as diffusion barrier with respect to the formation of hafnia at surface which can act as a passive layer when exposed in ablation environment. It has only limited industrial importance, possibly because of its high cost. Hafnium carbide powder is prepared by the reaction of HfO_2 with carbon at 1800-2200°C in hydrogen, a long processing time is required to remove all oxygen; by the carburization of hafnium sponge; by the carburization of hafnium hydride at 1600-1700°C. Hafnium carbide coatings are obtained by chemical vapor deposition (CVD) from a gas mixture of methane, hydrogen, and vaporized hafnium chloride [14-16].

In this chapter, we consider results of bulk, surface and interface properties of Hafnium carbide obtained from a number of density functional theory studies, including our own. Especially, the development of the methodology for the electronic structure calculation and hardware is remarkable in decades. In the past it was very difficult to perform the total energy pseudopotential calculation of transition metals, first period elements and related compounds. At present, however, this difficulty has been solved.

BULK PROPERTIES

Crystalline Structure

The equilibrium phase diagram of the Hf-C system [17-19] is shown in Figure 1. The nonstoichiometric carbide HfC_y exists in the disordered state in the Hf-C system, this compound has a relatively broad homogeneity interval and the NaCl type structure. An ordered phase of M_2C type is not formed in HfC_y, because its carbon content is smaller than those in disordered carbide HfC_y which corresponds to the lower boundary of the homogeneity interval. In addition, the ordered stoichiometric phases Hf_3C_2 and Hf_6C_5 form upon ordering under thermodynamic equilibrium conditions. The phase diagram of the Hf-C system also includes a new compound Hf_8C_{12}. This recently discovered compound belong to the metallocarbohedrenes group M_8C_{12} (M = Ti, Zr, Hf, V, Nb, Ta) [20].

Well known HfC_y crystallizes in the NaCl (rock-salt) type structure (space group $Fm\bar{3}m$) at any value of y. However, the ordered stoichiometric phases Hf_3C_2 and Hf_6C_5 should exist with possible space groups $Immm, P\bar{3}m1$, $P2$, or $C222_1$ for Hf_3C_2 and $P3_1, C2/m$, or $C2$ for Hf_6C_5, according to the results of the theoretical calculation by Gusev and Rempel [21].

Recently, Q. Zeng et al. [22] studied the stable compounds in the hafnium-carbon (Hf-C) system at ambient pressure using a variable-composition ab initio evolutionary algorithm and revealed that HfC, Hf_6C_5, and Hf_3C_2 are thermodynamically stable compounds; Whereas, Hf_2C is a metastable phase. As can be seen the most prominent stable state is HfC. Hf_3C_2 and, especially, Hf_6C_5 will only be stable in a narrow range of chemical potentials in hafnium-rich conditions. This result explains why HfC is well known from experiments, whereas, Hf_3C_2 and, especially, Hf_6C_5 are more elusive.

The calculations of the enthalpies of formation [22] indicate that the structure $C2/m$ is more stable with 20 atoms in the conventional unit cell of Hf_3C_2 and 22 atoms in the conventional unit cell of Hf_6C_5 [23]. The crystal structures of stable hafnium carbides (HfC, Hf_3C_2 and Hf_6C_5) are shown in Figure 2, and their crystallographic data are listed in Table 1.

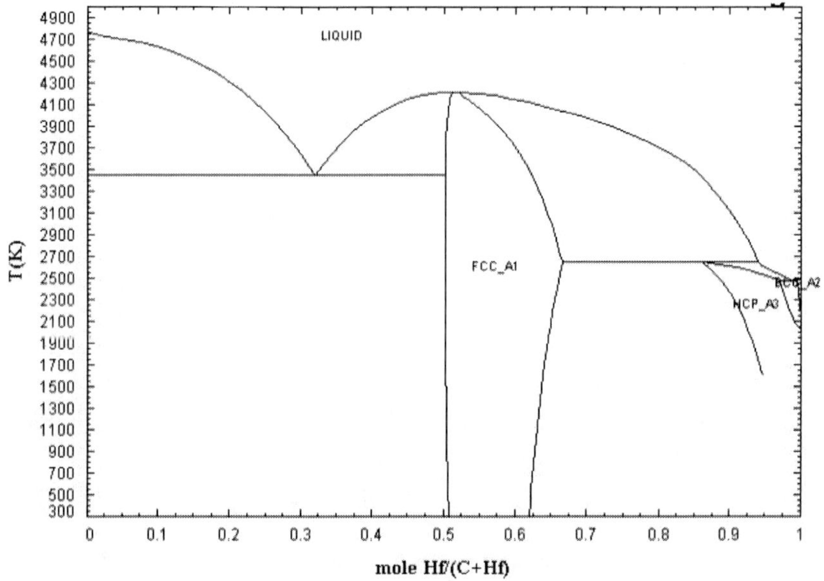

Figure 1. Equilibrium phase diagram of the Hf-C system.

Table 1. Crystallographic data of stable hafnium carbides

Compound	Space group	Atomic positions	Lattice constants (Å)
HfC	$Fm\bar{3}m$	Hf(*4a*) (0,0,0) C(*4b*) (0.5,0.5,0.5)	a = 4.64
Hf_3C_2	$C2/m$	Hf(*4i*) (0.737,0.5,0.744) Hf(*8j*) (0.248,0.161,0.264) C(*2a*) (0,0,0) C(*2d*) (0,0.5,0.5) C(*4h*) (0,0.834,0.5)	a = 5.642 b = 9.757 c = 5.636 β = 109.70°
Hf_6C_5	$C2/m$	Hf(*4i*) (0.739,0,0.260) Hf(*8j*) (0.241,0.327,0.746) C(*4g*) (0,0.333,0) C(*2d*) (0,0.5,0.5) C(*4h*) (0,0.832,0.5)	a = 5.656 b = 9.748 c = 5.657 β = 109.58°

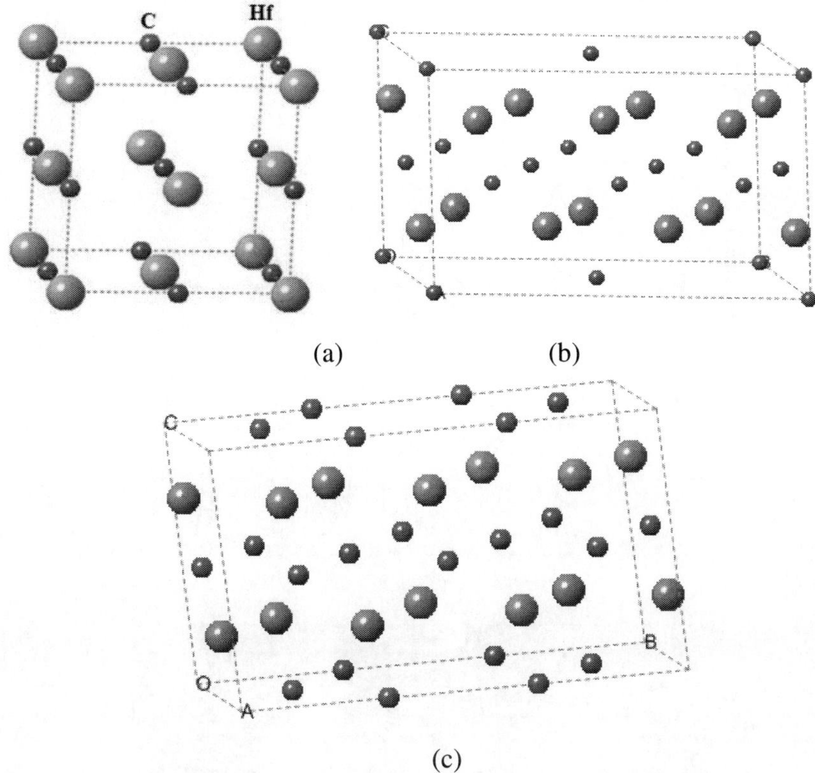

Figure 2. The crystal structure of (a) HfC, (b) Hf_3C_2, and (c) Hf_6C_5. the dark gray spheres present the C atoms, the yellow spheres present the Hf atoms.

ELECTRONIC PROPERTIES AND BONDING

Generally, the bonding in carbides arises from the interaction of carbon 2s and 2p orbitals with metal d orbitals. For HfC, Many first-principles density functional theory (DFT) calculations have been applied to study the electronic properties, including our own. We display in Figure 3 the electronic densities of states (DOS) and charge density contours on the (100) plan.

The DOS of HfC consists of two main regions: the first one is principally dominated by carbon 2s states and the second interval, constituting the bonding states below the Fermi level, is mainly composed of the hybridization of carbon 2p and hafnium 3d (4d) states. The finite DOS value at the Fermi level implies metallic nature, but its small value (nearly a pseudo-gap) suggests strong covalent character. Antibonding levels above the Fermi level are mainly the delocalized d metallic states.

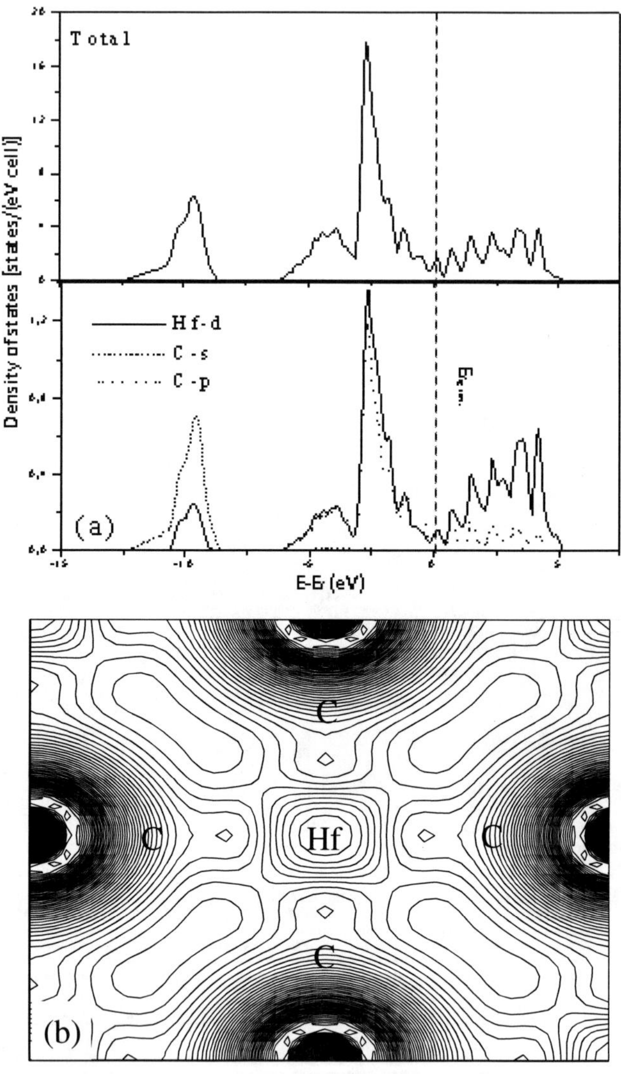

Figure 3. Electronic structure of bulk HfC: (a) Total and partial density of states (DOS), and (b) Charge density contours taken on the (100) plane.

The charge density contours show partial charge transfer from the hafnium atom to the carbon atom. This is consistent with the difference in electronegativity between Hf and C atoms and suggests some ionic character. The bonding in HfC exhibits then a mixture of metallic, ionic, and covalent character, which is consistent with other results on transition metal carbides:

Ti, V, Zr, Hf, and Nb [24-28]. For Hf_3C_2, and Hf_6C_5, the electronic structure and bonding are discussed by Q. Zeng and co-workers [22] and found the bonding in both compounds can be characterized as mixed metallic covalent due to the pronounced pseudogaps at the Fermi level and the strong hybridization of C-p and Hf-d valence states below the Fermi energy.

PHYSICAL AND MECHANICAL PROPERTIES

The hafnium carbide (HfC), like the group IV carbides, is well known for their hardness and strength, the Table 2 summarizes the characteristics and properties of HfC, compared with those of the pure Hf [29,30]. The physical and mechanical properties of HfC resemble more those of ceramics than those of metals. Where, the melting point of HfC is higher than of the pure Hf and the thermal conductivity is similar to that of the pure Hf. HfC is good electrical conductor and has an electrical resistivity only slightly higher than that of the Hf, reflecting the metallic character of this compound.

Table 2. Characteristics and Properties of HfC and Hf

Properties	HfC	Hf
Crystal Structure	B1 NaCl	Hcp
Lattice Parameter, Å	4.64	a = 3.19, c = 5.05
Space Group	Fm3m	p6₃/mmc
Pearson Symbol	cF8	hp1
Composition	$HfC_{0.60}$ to $HfC_{0.99}$	
Melting points, °C	3928	2233
Debye Temperature, K	436	252
Specific Heat (C_p), J/mole.K	37.4	26.27
Density, g/cm³	12.67	13.20
Thermal Conductivity, W/m.K	22.15	23
Coefficient of Thermal Expansion, 10^{-6}/K	6.3	6
Electrical Resistivity, μΩ.cm	37-45	30.6
Magnetic Susceptibility, 10^{-6} emu/mol	-23	
Vickers Hardness, GPa	26.1	1.76
Bulk Modulus, GPa	242	110
Shear Modulus, GPa	193	30
Young's modulus	461	78
Poisson's Ratio	0.18	0.37
Oxidation Resistance	Oxidizes in air at 500°C	

Table 2. (Continued)

Chemical Resistance	
HfC	Dissolved by cold HNO_3 and by a cold mixture of H_2SO_4 and H_3PO_4. Reacts readily with the halogens. Can be heated in hydrogen to its melting point without decomposition.
Hf	Hf is soluble in hydrofluoric and concentrated sulfuric acids and aqua regia. It begins to react slowly with air or oxygen at about 400°C and nitrogen at about 900°C and rapidly with hydrogen at about 700°C.
Isomorphism	
HfC	HfC forms solid solutions with oxygen and nitrogen which have a wide range of composition. HfC forms solid solutions with the other monocarbides of Group IV and V, particularly NbC and the solution HfC-NbC is used as coating for tools.
Hf	Commercial Zr metal typically contains 1–2.5% of Hf, which is not problematic because the chemical properties of Hf and Zr are similar. Their neutron-absorbing properties differ strongly, however, necessitating the separation for applications involving nuclear reactors.

Thus, Hafnium carbide combines the physical properties of three different classes of materials: covalent solids, ionic crystals, and transition metals. As a result, HfC often demonstrates the extreme hardness of covalent solids, the high melting temperature of ionic crystals, and the excellent electric and thermal conductivity of transition metals.

Elastic properties define the mechanical strength of materials and provide valuable information about the response of materials against the externally applied stress. In addition, they can also serve as a measure of phase stability of materials. Several experimental and theoretical studies [24-27,31-36] calculated the elastic constants for HfC, Table 3 summarizes some experimental and theoretical values of the modulus of elasticity.

The calculated elastic constants of HfC satisfy the stability conditions:

$$C_{11} + 2C_{12} > 0, \; C_{11} - C_{12} > 0, \; C_{44} > 0 \tag{1}$$

The small deviations may come from different calculation techniques or different experimental conditions, e.g., different package such as CASTEP, VASP, WIEN or ABINIT with LDA or GGA methods is used, and the calculated results at 0 K are comparing with those measured at room temperature.

Table 3: Calculated and experimental elastic constants Cij, bulk modulus B, shear modulus G, Young's modulus E (in GPa) and Poisson's ratio V of HfC

HfC	C_{11}	C_{12}	C_{44}	B	G	E	V
Exp	500[a]	114[b]	180[a]	242[b]	195[b]	461[b]	0.18[b]
PP	527.6[c]	110.6[c]	160[c]	249.6[c]	177.9[c]	431.2[c]	0.212[c]
FPLAPW	502[d]	97[d]	143[d]				

[a] Ref. [31]

[b] Ref. [32]

[c] Ref. [25]

[d] Ref; [24]

Q. Zeng et al. [22] calculated also the elastic constants of Hf_3C_2 and Hf_6C_5 at the ground state (listed in Table 4). The calculations of C_{ij} satisfy criteria of stability:

$$K_2 = \det|Cij|, \; i, j \leq 5, \; K_2 > 0, \; C_{44}C_{66} - C_{46}^2 > 0 \tag{2}$$

The results of elastic properties confirm that HfC, Hf_3C_2 and Hf_6C_5 are mechanically stable.

Table 4. Elastic properties (GPa) of Hf_3C_2 and Hf_6C_5 [22]

Compound	C_{11}	C_{22}	C_{33}	C_{44}	C_{55}	C_{66}	C_{12}	C_{13}	C_{16}
Hf_3C_2	386	404	362	125	116	131	106	113	-8.6
Hf_6C_5	429	462	460	184	162	185	128	130	4.2
	C_{23}	C_{26}	C_{36}	C_{45}	B	G	K	H	
Hf_3C_2	102	4.4	6.5	-9.1	199	129	0.65	17.74	
Hf_6C_5	110	26.7	-18.1	22.1	232	170	0.73	24.92	

HfC SURFACE PROPERTIES

Experimental and theoretical surface science studies have played an important role in confirming and understanding the surface properties of TMCs. TMC surfaces offer excellent model systems to determine how the electronic properties of metal surfaces are modified by the incorporation of carbon and how the electronic modifications in turn alter the surface chemical properties. For ideal surfaces of cubic crystals, the most commonly studied crystalline planes are the (100), (111) and (110) surfaces. In case of HfC (illustrated in Figure 4), the (100) and (110) surfaces are characterized by the coexistence of the hafnium and carbon atoms; each layers contains one Hf and one C. The (111) orientation has alternating layers of pure Hf and pure C atoms, and the (111) surface can be terminated either by hafnium or by carbon. L. I. Johansson [37] believed that the (100) surface has a low energy surface in many stable carbides. A. Arya and E. A. Carter [38,39] confirmed that stabilities of both ZrC and TiC surfaces follow the packing density sequence (110) < (111) < (100).

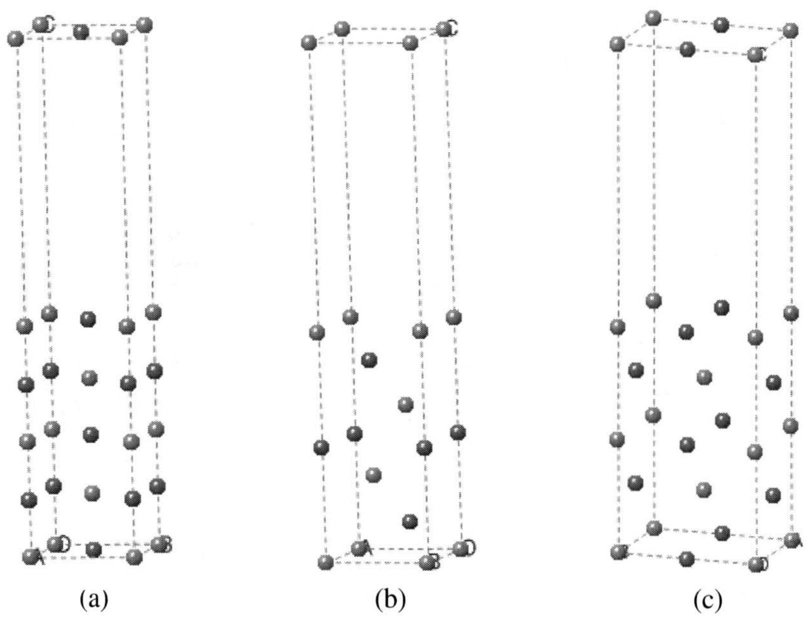

(a)   (b)   (c)

Figure 4. The surface structure of (a) HfC (100), (b) HfC (111) and (c) HfC (110).

The properties evaluated are the surface energies, relaxations, and electronic properties. The surface energy is conventionally defined as the reversible work that is required to build an unit area of a particular surface. In a slab model, it is usually calculated using [40]:

$$E_{surf} = \frac{E_{slab} - NE_{bulk}}{2S} \tag{3}$$

where E_{slab}, is the total energy of surface supercell slab, N is the total number of atoms in the slab, E_{bulk} is the total energy per each atom in the bulk material, S is the corresponding surface area, and the factor 2 accounts for the two identical surfaces of the slab.

Table 5 shows the calculated surface energies of the HfC (100) for slabs thicknesses ranging from 3 to 11 layers with two pseudopotentials Ultra-Soft (US) [41] and Projector Augmented Wave (PAW) [42]. The surface energy for HfC converges rapidly with increasing slab thickness to within about 0.01 J/m2 for slabs with n≥7. For two more planes of HfC (111) and (110), the surface energies are higher than that of HfC (100), which is, therefore, more stable.

Table 5. Convergence of the surface energy of HfC (100) with respect to the number of layers

Number of layers (n)	Surface energy of HfC (100)	
	PP-US	PP-PAW
3	1.85	1.75
5	1.81	1.72
7	1.78	1.70
9	1.77	1.70
11	1.76	1.70

Figure 5 shows the total and partial density of states (DOS) for the outermost layer of the HfC (100) surface. The DOS is composed of two regions: the first one ranging from -12 eV to -8 eV is dominated by carbon 2s states and the second interval ranging from -5 eV to Fermi level is composed of carbon 2p and hafnium 3d. Their overlap is very large, indicating that there exists strong hybridization between the C-2p and the Hf-3d.

A comparison with the bulk DOS (Fig. 3-a) reveals that new states appear in the energy range between -5 and 5 eV around the Fermi level. The C-2p and Hf-3d peaks in the occupied part of DOS shift towards lower binding energy by about 0.9 eV, thereby reducing the valence band width and enhancing the degree of localization of electrons at the surface.

Similarly, the low-lying C-2s peak is also shifted towards smaller binding energies by about 0.5 eV at the surface. This result consist with other calculated DOS of HfC (100) [43] and Group IV carbides TiC (100) [38], ZrC (100) [39].

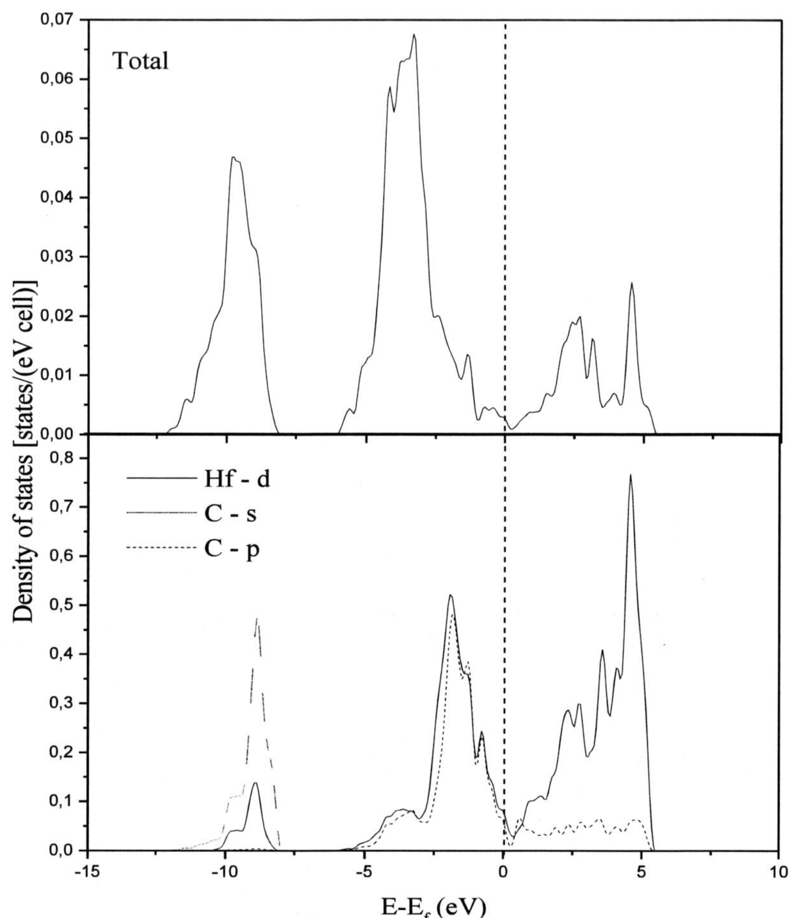

Figure 5. Total and partial density of states (DOS) for the top atomic layer of the HfC (100) surface.

METAL/HfC INTERFACES

Under the high temperature and pressure conditions present in many industrial applications, coatings must be used to protect substrate from thermomechanical failure. The successful protective coating includes high melting point, wear resistance, corrosion resistance, and strong adhesion to the substrate. Since chrome coatings satisfy many of these requirements, due to their properties such as the strong adhesive ability with substrate, high hardness, excellent wear and corrosion resistance, but the chrome coatings fail when inherent microcracks allow deleterious gases to penetrate and erode the Cr/M interface [44,45]. Consequently, a new protective coating system has long been desired [46]. Alloying has not proven successful in protecting, for example protecting steel in high-temperature environments (stainless steel Cr-doped offers protection against corrosion, but its melting point is too low). For this reason, the researchers have turned their attention to exploring alternative protective coatings based on ceramic materials rather than metal alloys. The most commonly used ceramic particles for reinforcement of various types in metal matrix composites include various oxides (e.g., Al_2O_3, Y_2O_3 and ZrO_2), nitrides (e.g., TiN and Si_3N_4), and carbides (e.g., SiC, TiC, Cr_3C_2, VC, WC and B_4C). These particulates are generally used to increase wear resistance. HfC is one of the most optimal protective coating materials due to their physical and mechanical properties and your ability to be entirely wetted by substrate e.g., iron-group metals [47].

In this section, we expose our results of the First-principles calculations of the Mo(110)/HfC(100), Fe(110)/HfC(100), Co(111)/HfC(100) and Ni(111)/HfC(100) interfaces [48-50].

Interface Geometry

Generally speaking, there are a huge number of ways two surfaces can be joined to form an interface: the surfaces can be created by cleaving along one of many possible planes, when dealing with compounds one has to choose among several surface stoichiometries, and finally there is a continuum of relative rotational and translational orientations. However, as indicated by Christensen and co-workers, [51,52] the stable interfaces are generally formed between most stable-free surfaces. Thus, we restricted our study to the Mo (110)/HfC (100), Fe (110)/HfC (100), Co (111)/HfC (100) and Ni (111)/HfC (100) interfaces.

Based on the convergence of the surface energies, the interfaces model uses seven layers of HfC(100) placed on nine layers of Mo (110), nine layers of Fe(110), seven layers of Co(111), and seven layers of Ni(111).

To identify the optimal interface geometry we considered two different stacking sequences, placing the interfacial substrate M in one of two positions with respect to the HfC surface lattice structure: M above the C atoms (C-site), M above the Hf atoms (Hf-site). As illustrated in Figure 6, the interface models for C- and Hf-site (Fe, Co, Ni)/HfC.

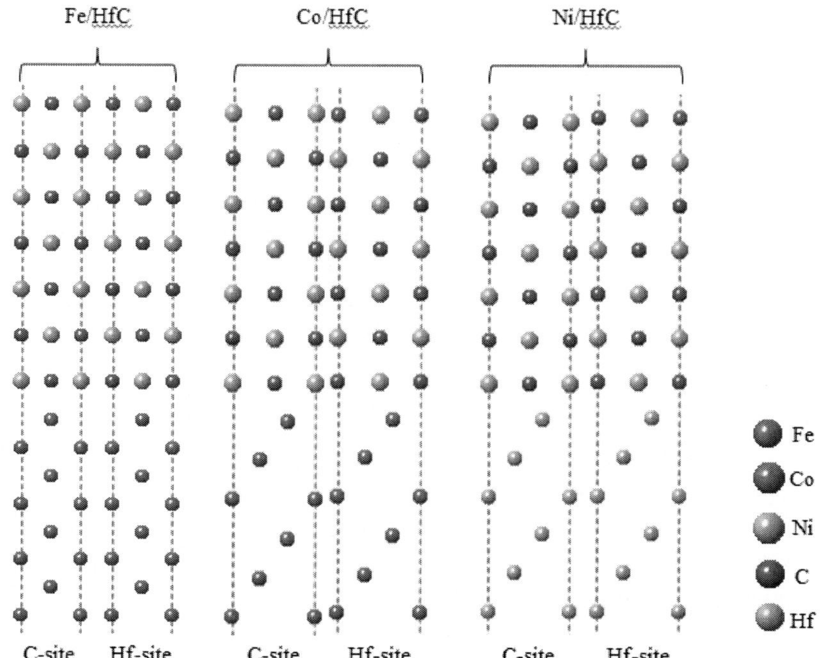

Figure 6. The interface models for C- and Hf-site M (Fe, Co, Ni)/HfC, where M above the C atoms (C-site) and M above the Hf atoms (Hf-site).

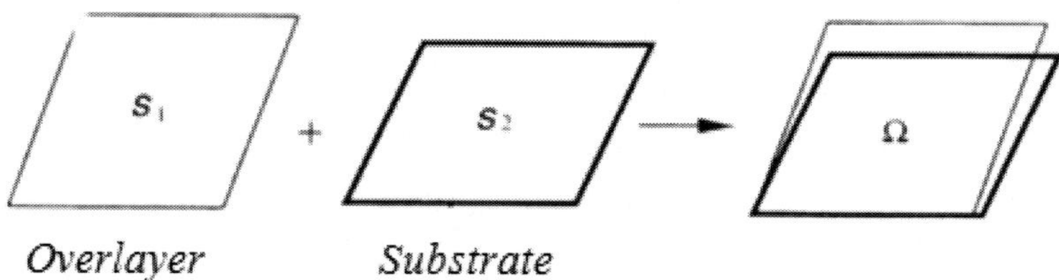

Figure 7. Schematic picture of M/HfC interface lattice matching. S_1 and S_2 are the surface unit cell of HfC and M, respectively. Ω is the overlap area between the cells when overlaid.

To compensate the mismatch, the softer M matches the lattice of HfC. The interface matching is sketched in Figure 7: some surface unit cell of substrate with area S_2 is distorted to fit with the HfC surface unit cell with area S_1, and an overlap area Ω is then calculated. The misfit parameter ξ defined as [52]:

$$\xi = \frac{2\Omega}{S_1 + S_2} \tag{4}$$

Our calculated value is about 1.8%, 1.15% and 1.02% for Fe/HfC, Co/HfC and Ni/HfC, respectively. Full relaxation will allow the atoms to move to more favorable positions and the atomic structure near the interface, predominantly that of the elastically softer material, will relax so as to lower the interfacial energy.

Interface Adhesion

An important property of the interface is its excess free energy per unit area and the closely related work of adhesion W_{ad}. Thermodynamic and mechanical properties of the interface have been found to depend on these parameters. W_{ad} is defined as the bond energy needed (per unit area) to reversibly separate an interface into two free surfaces, neglecting plasticity [53,54]. The value of W_{ad} is the lower limit on the work needed for a real cleavage experiment [55]. Formally, it can be given by the difference in total energy between the interface and its isolated surfaces:

$$W_{ad} = \frac{E_M + E_{HfC} - E_{M/HfC}}{2A} \tag{5}$$

where $E_{M/HfC}$ is the total energy of supercell containing the multi-layered slabs; E_M and E_{HfC} are the total energies of the same supercell containing a single slab of the substrate or HfC, respectively; A is the interface area and the factor 2 takes into account the presence of the two equivalent interfaces within a unit cell.

The calculated works of adhesion (W_{ad}) and interfacial separation (d_0) for all interfaces are listed in Table 6. In all cases, the interface formed with M atoms placed above the C atoms exhibits larger work of adhesion: 3.30 J/m² for Mo/HfC, 2.38 J/m² for Fe/HfC, 2.78 J/m² for Co/HfC and 3.02 J/m² for Ni/HfC. Compared to the C-site, the work of adhesion for Hf-site is significantly weaker: 1.19 J/m², 1.59 J/m², 1.96 J/m² and 2.37J/m² for Mo/HfC, Fe/HfC, Co/HfC and Ni/HfC interfaces, respectively.

In the case of iron-group metals (Fe, Co, Ni)/HfC interfaces, we also note that the W_{ad} increases on passing from iron to nickel ($W_{ad(Fe/HfC)} < W_{ad(Co/HfC)} < W_{ad(Ni/HfC)}$), this result can be explained by considering the model of configurational localization of valence electrons [56]. As transition metals of Sub-group IVa are donors of electrons (the unlocalized valence electrons) and the iron-group metals are electron acceptors, whose accepting capacity grows in the order Fe < Co < Ni. Thus, in systems composed of an iron-group metal and a refractory carbide (especially HfC), the presence of a metallic acceptor and a carbide donor ensures good adhesion between the two components (the adhesion increases in the order Fe < Co < Ni).

Table 6. Calculated relaxed work of adhesion (W_{ad}) and interfacial separation (d_0) for the M/HfC interfaces

M/HfC interface	Stacking	d_0 (Å)	W_{ad} (J/m^2)
Mo/HfC	C-site	1.68	3.30
	Hf-site	2.49	1.19
Fe/HfC	C-site	2.02	2.38
	Hf-site	2.41	1.59
Co/HfC	C-site	2.34	2.78
	Hf-site	2.94	1.96
Ni/HfC	C-site	2.51	3.02
	Hf-site	3.47	2.37

Interface Bonding

The mechanical properties of an interface are closely related to the interfacial atomic bonding. The valence electron charge density differences of the interfaces are given in Figure 8.

The valence electron charge density difference $\Delta\rho$ is given by:

$$\Delta\rho = \rho_{M/HfC} - \rho_M - \rho_{HfC} \tag{6}$$

where $\rho_{M/HfC}$ is the valence charge density of the interface system; ρ_M and ρ_{HfC} are calculated for isolated substrate M and HfC slabs of the same supercell, respectively.

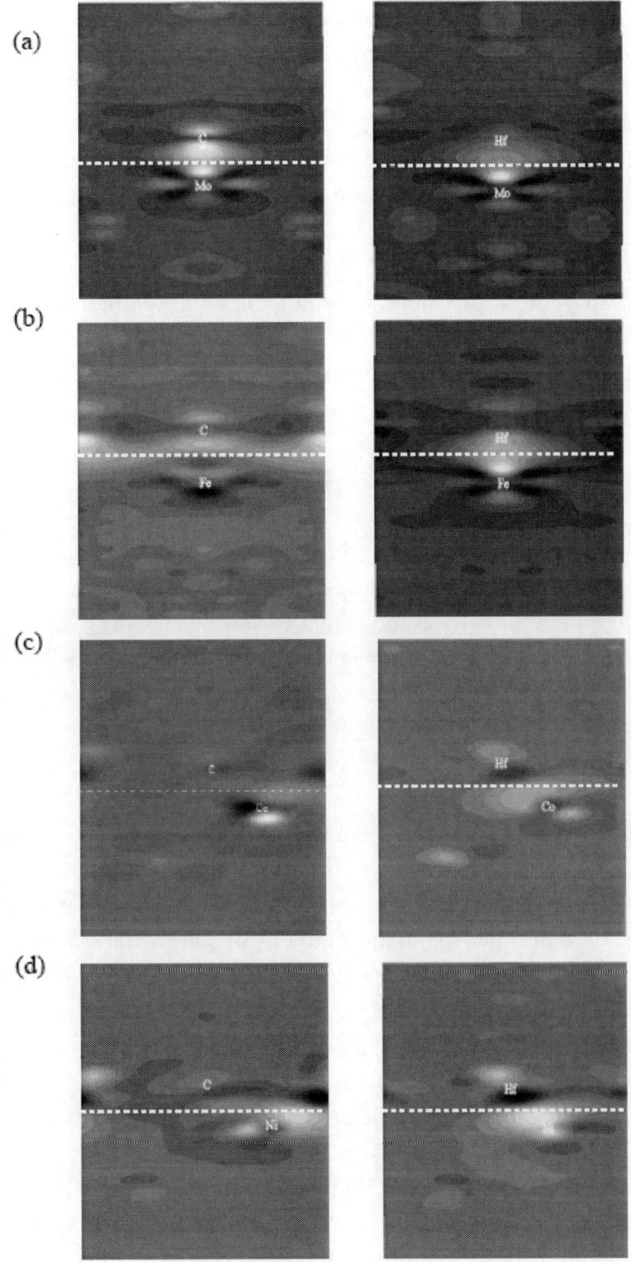

Figure 8. Charge density difference taken along the (010) plane for: (a) C- and Hf-site Mo/HfC, (b) C- and Hf-site Fe/HfC, (c) C- and Hf-site Co/HfC and (d) C- and Hf-site Ni/HfC interfaces.

For C-site interface, the interfacial M atom is dragged very near to the interfacial C atom, where the charge accumulation is slightly closer to the C atom than to the M one. This is consistent with the difference in electronegativity between M and C atoms. This charge accumulation is typically characteristic of a polar covalent bonding. Accordingly, the larger work of adhesion corresponding to the C-site in all interfaces can be directly related to the interfacial bonding mechanism and the polar covalent/ionic mixed bonds found between the M and C atoms.

For Hf-site interface, there is small charge accumulation between the interfacial M and Hf atoms, which is characteristic of weak covalent bonding. Thus the bonding at the interface remains mainly metallic.

Interface Magnetism

The calculated magnetic moments of the atoms in each layer of the Fe/HfC interface for the C- and Hf-site are shown in Figure 9.

The magnetic moments at the center of the Fe slab are 2.211 μ_B for C-site and 2.192 μ_B for Hf-site and at the center of the HfC slab it is 0 μ_B. This close agreement to bulk magnetic moments indicates that the magnetic properties are converged with respect to slab thickness.

For C-site, the magnetic moment of interfacial Fe (2.335 μ_B) is higher than the bulk value of 2.2 μ_B and the Fe atoms at the interface induce a magnetization in the interfacial C atoms (-0.204 μ_B). Moreover, the interfacial Fe atoms have reduced magnetic moments compared to their surface value of 2.486 μ_B, suggesting strong spin pairing is involved in Fe–C interactions. Similarly for Hf-site, the magnetic moment of interfacial Fe is 2.281 μ_B and the Fe atoms at the interface induce a small magnetization in the interfacial Hf atoms (-0.094 μ_B). Thus, the spin pairing at these interfaces is likely to be the origin of the enhanced bonding compared to pure Fe.

The increase of the Fe magnetic moment at the Fe/HfC interface can be explained by the change in their electronic structure. This is clearly reflected in the local density of states (LDOS) plotted for C-site and Hf-site interfaces and compared to the Fe bulk states (Figure 10). The Fe LDOS, for the band of majority-spin of C-site (Figure 10-a) and Hf-site (Figure 10-b) is strongly different from the bulk one (Figure 10-c). It is shifted to lower energies so that the number of electrons, of spin up, accommodated below the Fermi level is higher. The Fe LDOS for the minority spin is not essentially modified. As the consequence of these modifications i.e., an increase (decrease) of the electrons of spin up (spin down), the magnetic moment on the Fe atom at the interface went up from 2.2 μ_B to 2.335 μ_B for C-site and from 2.2 μ_B to 2.281 μ_B for Hf-site.

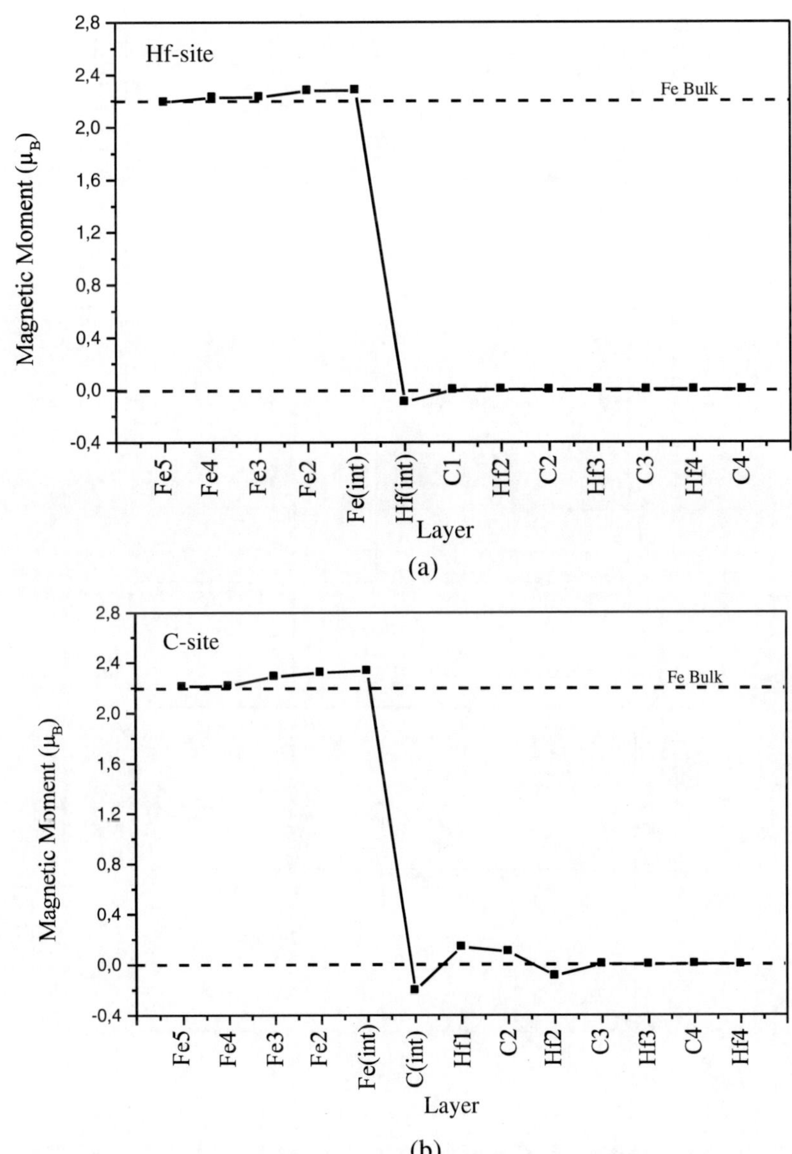

Figure 9. Magnetic moments per atom for Fe/HfC interface: (a) C-site and (b) Hf-site.

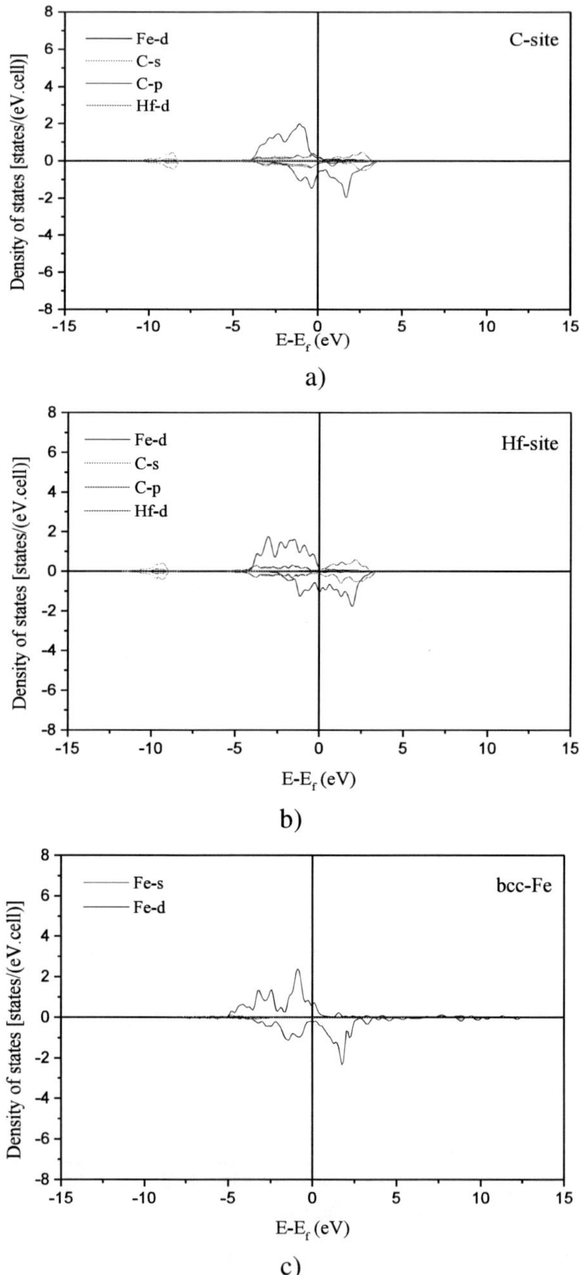

Figure 10. Layer-projected partial DOS for atoms nearest to the: (a) C-site and (b) Hf-site.

Contrary to the interfacial Fe where the spin up is strongly modified, it is the minority spin LDOS of the interfacial HfC which is strongly perturbed as compared to the corresponding DOS of bulk HfC (Figure 3-a). This gives rise to a smaller magnetic moment on the Hf and C atoms at the Fe/HfC interface.

We note the same effect for Co/HfC and Ni/HfC interfaces, the details in Ref. [50].

Effects of Alloying on Adhesion

It is known that impurities could greatly affect bonding and cohesion at imperfections, such as grain boundaries. Experimental studies [57,58] suggested that impurities also have substantial effects on wetting and adhesion at metal/ceramic interfaces through selective segregation of the alloying element at the interface.

We performed ab initio calculations in order to determine how Re affects the adhesion of Mo/HfC interface. We use the preceding supercells, assuming the Re atom substitutes one interfacial Mo atom of the most optimal C-site interface, giving a concentration of 50% Re.

We find that the work of adhesion is 3.85 J/m²; then, the presence of Re increases the work of adhesion: 0.55 J/m^2 (17%) for Mo/HfC. This implies that the impurity Re atom improves the adhesion. This is in good agreement with the available experimental results, which show that the addition of 40% Re into liquid Mo improves the wettability of molybdenum on HfC [59]. Previous studies [60,61] showed that the strain plays an important role in determining the effects of additional elements on the work of adhesion. Meanwhile, Mo and Re are of almost the same atomic size and so induce no considerable lattice strain when alloying. The presence of Re does not change d_0 for the Mo/HfC interface. Further, our analysis of the charge density and density of states shows that the impurity Re does not affect the nature and strength of the interface Mo-C bonds, which agrees with the results of other earlier studies, where the impurity atoms could not improve the bonding of the metal/ceramic interfaces [61-63].

CONCLUSION

Hafnium carbide offer potential as protective coatings for substrates because of its favorable physical, chemical, and mechanical properties. In this chapter, we have focused the understanding the behavior of the bulk, surface and interface formed with metal or alloy substrates.

Hafnium carbides only have three thermodynamically stable polymorphs at ambient pressure. In addition to the well-known HfC, Q. Zeng et al. [22] predicted two additional thermodynamically stable compounds Hf_3C_2 and Hf_6C_5. Hf_2C is a metastable phase. The structure of HfC with cubic symmetry ($Fm\bar{3}m$) contains 8 atoms in the conventional cell, whereas the stable structure of Hf_3C_2 with space group $C2/m$ contains 20 atoms in the conventional cell. The structure of Hf_6C_5 also has space group $C2/m$, which contains 22 atoms in the conventional cell. The electronic properties and bonding behaviors of Hafnium carbides are discussed by their density of states, electron charge density.

Strong covalent Hf-C bonding is noticed in all these weakly metallic compounds. DOS shows lower metallicity of HfC compared to that of Hf_6C_5 and Hf_3C_2. Thus, HfC could have the highest hardness and melting point among these three compounds. The results of elastic properties confirm the mechanical stability of HfC, Hf_3C_2 and Hf_6C_5.

For HfC surface, the stability was found to increase with packing density, with (100) most stable. HfC surface exhibited predominantly polar covalent bonding arising from Hf d-C p mixing. Based on this surface and other calculated substrate surfaces, Mo(110)/HfC(100), Fe(110)/HfC(100), Co(111)/HfC(100) and Ni(111)/HfC(100) interfaces are studied. Two different stacking sequences C- and Hf-site were considered, the calculations of work of adhesion declare that substrate bind preferentially with C atoms. The electronic structures of interfaces are analyzed using charge density differences; the results indicate that the bonding behavior of C-site interface is polar covalent bond in the interfacial region. While for the Hf-site interface, the interfacial Fe–Hf bond is mainly metallic.

The magnetic properties at the interfaces are calculated using spin polarized DOS, the results show that the formation of interface will enhance the magnetic moment of interfacial iron-group metals (Fe, Co and Ni) atoms for all C-site and Hf-site interfaces.

This enhancement is related to the change in band of majority-spin of iron-group metals. In addition, the effects of alloying elements at the interfaces were examined, and found that Re alloying at the Mo/HfC interface improves the adhesion, increasing W_{ad} by 17% but does not noticeably affect the interfacial bonding.

REFERENCES

[1] Jones, R. H. (2001). *Environmental Effects on Engineered Materials.* CRC Press.

[2] Li, H.; Zhang, L. T.; Zeng, Q. F.; Ren, H. T.; Guan, K.; Liu, Q. M.; Cheng, L.F. (2011) *Solid State Commun.*, 151, 61.

[3] Levine, S. R.; Opila, E. J.; Halbig, M. C.; Kiser, J.D.; Singh, M.; Salem, J. A. (2002) *J. Eur.Ceram. Soc.*, 22, 2757.

[4] Opeka, M.M.; Talmy, I.G.; Zaykosk, J.A. (2004) *J. Mater. Sci.*, 39, 5887.

[5] Savino, R.; Fumo, M.D.S.; Paterna, D.; Serpico, M. (2005) *Aerosp. Sci. Technol.*, 9, 151.

[6] Isaev, E. I.; Ahuja, R.; Simak, S. I.; Lichtenstein, A. I.; Vekilov, Y. K.; Johansson, B.; Abrikosov, I. A. (2005) *Phys. Rev.* B, 72, 064515.

[7] Isaev, E. I.; Simak, S. I.; Abrikosov, I. A.; Ahuja, R.; Vekilov, Y. K.; Katsnelson, M. I.; Lichtenstein, A. I.; Johansson, B. (2007) *J. Appl. Phys.*, 101, 123519.

[8] Pintschovius, L.; Reichardt, W.; Scheerer, B. (1978) *J. Phys. C*, 11, 1557.

[9] López-de-la-Torre, L.; Winkler, B.; Schreuer, J.; Knorr, K.; Avalos-Borja, M. (2005) *Solid State Commun.*, 134, 245.

[10] Chen, Z. W.; Gu, M. X.; Sun, C. Q.; Zhang, X. Y. ; Liu, R. P. (2007) *Appl. Phys. Lett.*, 91, 061905.

[11] Sayir, A. J. (2004) Mater. Sci., 39, 5995.

[12] Wang, Y. L.; Xiong; X.; Li, G. D.; Zhao, X. J.; Chen, Z. K.; Sun; W. (2012) Surf. Coat. Technol., 206, 2825.

[13] Krajewski, A.; D'Alessio, L.; De Maria, G. (1998) *Cryst. Res. Technol.*, 33, 341.

[14] Ferro, D.; Barinov, S. M.; Rau, J. V.; Latini, A.; Scandurra, R.; Brunetti, B. (2006) *Surf. Coat. Technol.*, 200, 4701.

[15] Teghil, R.; Santagata, A.; Zaccagnino, M.; Barinov, S. M.; Marotta, V.; De Maria, G. (2002) *Surf. Coat. Technol.*,151-152, 531.

[16] Li, G., Li, G. (2010) *J. Coat. Technol. Res.*, 7, 403.

[17] Gogotsi, Y. G.; Andrievski, R. A. (1999). *Materials Science of Carbides, Nitrides and Borides.*

[18] Gusev, A. I.; Zyryanova, A.N. (2000) Phys. stat. sol. (a), 177, 419.

[19] Gusev, A. I.; Rempel, A. A.; Magerl, A. J. (2001). *Disorder and Order in Strongly Nonstoichiometric Compounds.* Springer.

[20] Guo, B. C.; Wei, S.; Purnell, J.; Buzza, S.; Castleman, W. (1992) *Science,* 256, 515.

[21] Gusev, A. I.; Rempel, A. A. (1993) *Phys. stat. sol.* (a), 135, 15.

[22] Zeng, Q.; Peng, J.; Oganov, A. R.; Zhu, Q.; Xie, C.; Zhang, X.; Dong, D.; Zhang, L.; Cheng, L.; (2013) *Phys. Rev. B*, 88, 214107.

[23] Wyckoff, R.W.G. (1948). The Structure of Crystals. *Interscience*, New York.

[24] Zaoui, A.; Bouhafs, B.; Ruterana, P. (2005), *Mater. Chem. Phys.*, 91, 108.

[25] Li, H.; Zhang, L.; Zeng, Q.; Guan, K.; Li, K.; Ren, H.; Liu, S.; Cheng, L. (2011), *Solid State Commun.*, 151, 602.

[26] Yang, J.; Gao, F. (2012), *Physica B*, 407, 3527.

[27] Chauhan, M.; Gupta, D. C. (2013), *Diamond & Related Materials*, 40, 96.

[28] Siegel, D. J.; Hector Jr., L. G.; Adams, J. B. (2002), *Acta. Mater.*50, 619.

[29] Toth, L. E. (1971). *Transition Metal Carbides and Nitrides*. Academic, New York.

[30] Kittel, C. (1996). *Introduction to Solid State Physics*, 7th ed. Wiley, New York.

[31] Weber, W. (1973) *Phys. Rev. B,* 8, 5082.

[32] Brown, H. L.; Armstrong, P. E.; Kempter, C. P. (1966) *J. Chem. Phys.*, 45, 547.

[33] He, L. F.; Lin, Z. J.; Wang, J. Y.; Bao, Y. W.; Zhou, Y. C. (2008) *Scripta Mater.* 58, 679.

[34] Liu, Y.; Jiang, Y.; Zhou, R.; Feng, J. (2014) *J. Alloy. Comp.*, 582, 500.

[35] Lu, X. G.; Selleby, M.; Sundman, B. (2007) *Acta Mater.*, 55, 1215.

[36] Sai Gautam, G.; Hari Kumar, K. C. (2014) *J. Alloy. Comp.*, 587, 380.

[37] Johansson, L. I. (1995) *Surf. Sci. Rep.*, 21, 177.

[38] Arya, A.; Carter, E.A. (2003) *J. Chem.Phys.*,118, 8982.

[39] Arya, A.; Carter, E.A. (2004) *Surf. Sci.*, 560,103.

[40] Hugosson, H. W.; Eriksson, O.; Jansson, U.; Ruban, A. V.; Souvatzis, P.; Abrikosov, I. A. (2004) *Surf Sci,* 557, 243.

[41] Vanderbilt, D. (1990) *Phys. Rev. B,* 41, 7892.

[42] Blöchl, P. E. (1994) *Phys Rev B*, 50,17953.

[43] Liu, D.; Deng, J.; Jin, Y. (2010) *Comput..Mater..Sci.,* 47, 625.

[44] Turley, D. M. (1989) *Wear,* 131, 135.

[45] Cote, P. J.; Rickard, C. (2000) *Wear,* 241, 17.

[46] Carter, R. H. J. (2006) *Pressure Vessel Technol-Trans Asme*, 128, 251.

[47] Samsonov, G. V.; Panasyuk, A. D.; Kozina, G. K. D'yakonova, L. V. (1972) *Powder Metallurgy and Metal Ceramics*, 11, 568.

[48] Si Abdelkader, H.; Faraoun, H. I.; Esling, C. (2011) *J. Appl. Phys.*, 110, 044901.

[49] Si Abdelkader, H.; Faraoun, H. I. (2012) *J. Magn. Magn. Mater.*, 324, 4155.

[50] Si Abdelkader, H.; Faraoun, H. I. (2014) *J. Mater. Sci.*, 49, 407.

[51] Christensen, A.; Carter, E.A. (2000) *Phys. Rev. B*, 62, 16968.

[52] Christensen, A., Jarvis, E. A. A.; Carter, E. A. (2001). Atomic-level properties of thermal barrier coatings: characterization of metal-ceramic interfaces, Chemical Dynamics in Extreme Environments, *Advanced Series in Physical Chemistry:* edited by R. A. Dressler, series edited by C. Y. Ng, *World Scientific*, Singapore,Vol.11, p.490.

[53] Lipkin, D. M.; Clarke, D. R.; Evans, A. G. (1998) *Acta Mater.,* 46, 4835.

[54] Kriese, M. D.; Moody, N. R.; Gerberich, W. W. (1998) *Acta Mater.*, 46, 6623.

[55] Dudiy, S. V. (2002) *Surf. Sci.*, 497, 171.

[56] Samsonov, G. V. (1995) *Refractory Carbides*, Springer-Verlag US, p 1-11.

[57] Gibbesch, G.; Elssner, G. (1992) *Acta Metall. Mater.*, 40, S59.

[58] Korn, D.; Elssner, G.; Fischmeister, H. F.; Rühle, M. (1992) *Acta Metall. Mater.*, 40, S355.

[59] Manning, C. R.; Stoops, J. R. R. F. (1968) *J. Am. Ceram. Soc.*, 51, 415.

[60] Liu, L. M.; Wang, S. Q.; Ye, H. Q. (2004) *Surf. Sci.*550, 46.

[61] Siegel, D. J.; Hector Jr, L. G.; Adams, J. B. (2002) *Surf. Sci.*, 498, 321.

[62] Wang, X. G.; Smith, J. R.; Scheffler, M. (2002) *Phys. Rev. B*, 66, 073411.

[63] Wang, X. G.; Smith, J. R.; Evans, A. (2002) *Phys. Rev. Lett.*, 89, 286102.

In: Hafnium
Editor: HongYu Yu

ISBN: 978-1-63463-164-8
© 2015 Nova Science Publishers, Inc.

Chapter 4

STABILIZATION OF HIGHER SYMMETRY HfO$_2$ POLYMORPHS AS THIN FILMS AND NANOPARTICLES

Protima Rauwel[1] and Erwan Rauwel[2]*

[1]Department of Physics, University of Tartu, Tartu, Estonia
[2]Tartu College, Tallinn University of Technology, Tartu, Estonia

ABSTRACT

In the beginning of 21st century, hafnium dioxide (HfO$_2$) and HfO$_2$-based compounds were considered to be promising candidates for the replacement of SiO$_2$ as gate dielectric oxides for advanced Complementary Metal-Oxide Semiconductor (CMOS) technology and future advanced microelectronic applications. Numerous investigations and studies on the deposition of HfO$_2$ in the thin film form and on the improvement of its physical properties were then conducted, aiming more particularly at the increase in its permittivity. In fact, theoretical calculations had predicted that HfO$_2$ polymorphs with higher symmetry exhibit a higher permittivity than the thermodynamically stable monoclinic phase at room temperature. However, these higher symmetry phases are only stable at high temperatures.

The stabilization of these higher symmetry polymorphs can be brought about by two methods: a substrate induced strain induced by a difference of expansion coefficients between the HfO$_2$ thin film and the

---

* Corresponding author address: Email: protima.rauwel@ut.ee, erwan.rauwel@ttu.ee

substrate or by an oxygen sub-stoichiometry in the HfO_2 structure. It is well understood that the existence of oxygen vacancies in the HfO_2 thin films leads to the stabilization of its higher symmetry phases. In this chapter, the stabilization of higher symmetry phase of HfO_2 in the thin film form and nanoparticle form will be discussed. The possibility of stabilizing these polymorphs using substrate induced strains, thermal treatment and doping will be reviewed.

Keywords: HfO_2, polymorphs, thin films, nanoparticles, oxygen vacancies, doping, annealing

INTRODUCTION

HfO_2 compounds and their solid solution are recognized as technologically important materials [1] and more particularly since 2000s as a possible candidate for the replacement of SiO_2 gate oxides. For more than four decades, SiO_2 was the material of choice for gate oxides in Si field effect transistor (FET) technology. In the beginning of 21st century, the continuous scaling down of Si-based gate dielectrics was reaching its critical limit, leading to unacceptably high leakage currents due to direct tunneling and poorer reliability. In order to reduce leakage currents and improve reliability, high-κ dielectrics were investigated as alternatives to the SiO_2 gate dielectric [2]. Among the suggested materials, hafnium dioxide (HfO_2) and HfO_2-based compounds were considered as a serious alternative [3] for gate dielectric oxides for advanced Complementary Metal-Oxide Semiconductor (CMOS) technology and future advanced microelectronic applications due to its high dielectric constant [4], wide band gap ($\cong 5.68eV$) with suitable band offsets on silicon, high interface quality, low densities of interface states and bulk charges, a high thermal stability and a good ability to control the gate threshold voltage [5, 6]. These qualities have spurred numerous investigations and studies on the deposition of HfO_2 in the thin film form and on the improvement of its physical properties, more particularly in the increase of the permittivity for various microelectronic applications like CMOS technology [5, 7], MIM capacitors [8], and flash memory applications [9]. In fact, theoretical calculations have predicted that HfO_2 polymorphs with higher symmetry exhibit a higher permittivity than the thermodynamically stable monoclinic phase at room temperature. Values of 26-29 for the cubic phase and of 33-70 for the tetragonal phase, have been calculated which are higher than the value of 16-19 for the monoclinic phase. It is therefore desirable to stabilize these higher-symmetry phases. However, these phases are stable at only high temperatures. In the bulk form, the monoclinic to tetragonal phase transition occurs at 1700°C and the tetragonal-to-cubic phase transition occurs at 2700°C [10].

The stabilization of these polymorphs presenting a higher symmetry can be brought about by two methods. First, a substrate induced strain induced by a difference of expansion coefficients between the HfO$_2$ thin film and the substrate on which the film is deposited. This implies that a thermal expansion will induce a compressive substrate-induced biaxial strain on the film which will then stabilize the lower unit cell volume of the orthorhombic or cubic phase. In the beginning of the thin film growth, formation of a mixed orthorhombic, tetragonal, cubic and monoclinic structure is usually observed. The second method is an oxygen sub-stoichiometry in the HfO$_2$ structure, the existence of oxygen vacancies in the HfO$_2$ thin films will lead to the stabilization of the higher symmetry phases. HfO$_2$, like ZrO$_2$, is a fluorite-type structure, and similarly to ZrO$_2$, the cubic phase can be stabilized by adding an appropriate divalent or trivalent dopant. Diverse oxides such as CaO, MgO, SrO, SiO$_2$, CeO$_2$, TiO$_2$, Sc$_2$O$_3$ and Y$_2$O$_3$ and most of the rare-earth oxides have been studied for the possible stabilization of the cubic structure of HfO$_2$. Another possibility is, instead of incorporating oxygen vacancies by using a trivalent or divalent dopant, to use a reductive solvent during the growth of HfO$_2$ that will promote oxygen vacancies in the structure and, in consequence, will stabilize a higher symmetry phase.

In this chapter, the stabilization of higher symmetry phase of HfO$_2$ in the thin film form and nanoparticle form will be more particularly discussed. The possibility of stabilizing these polymorphs using substrate induced strains, thermal treatment and doping will be reviewed in the case of thin film growth and nanoparticle synthesis.

HfO$_2$ COMPOUND

HfO$_2$ in the Bulk Form

The Hf element was discovered in 1922 and it was found that the chemical behavior of this element was very similar to Zr, despite differences in atomic weight and density [11]. In fact, both materials present similar structures in their oxide form, but at different temperatures. There was also a noticeably different reactivity of these 2 elements and the formation of different solid solutions when mixed with calcium, carbon and more particularly silicon. In addition, it was observed that the thermal stability of $HfO_2/HfSiO_4$ is better compared to $ZrO_2/ZrSiO_4$, especially when the films were subjected to annealing temperatures under 1000°C in O_2 or NH_3 thereby proving that they do not naturally precipitate silicate compounds at this temperature [12]. HfO_2 compound is of technological importance for its extraordinary high Young's modulus [13], very high melting point, high chemical stability, as well as its high neutron absorption cross section [14]. More recently, hafnium based compounds have be recognized as the designated dielectrics for the future advanced microelectronic applications and more particularly MOSFETs. HfO_2 compound has three natural polymorphs; monoclinic, tetragonal and cubic polymorph. In the bulk form, HfO_2 naturally possesses the monoclinic baddeleyite polymorph ($P2_1/c$) that is stable at atmospheric pressure and temperature lower than 1700°C [15]. The monoclinic structure was for the first time accurately observed by Adam et al. [16], where they compared the crystallographic structures of HfO_2 and ZrO_2. The monoclinic structure results from a seven-fold coordination of Hf^{4+} and has an asymmetric unit cell composed of one Hf atom and two oxygen atoms, coordinated differently to the Hf atom. The simultaneous presence of 3 fold coordinated O indicated as "3" in Figure 1 and 4-fold coordinated O indicated as "4" play an important role in the electronic structure and induces important differences compared to the cubic polymorph in which only 4-fold coordinated O atoms are present.

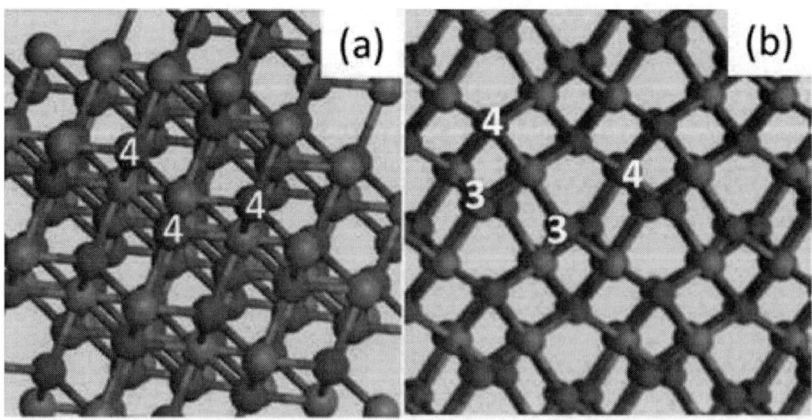

Figure 1. (a) Tetrahedrally coordinated O in the cubic structure, (b) 3-fold and 4-fold coordinated O in the monoclinic structure of HfO_2. Reprinted from reference [19] copyright license 3404760357695.

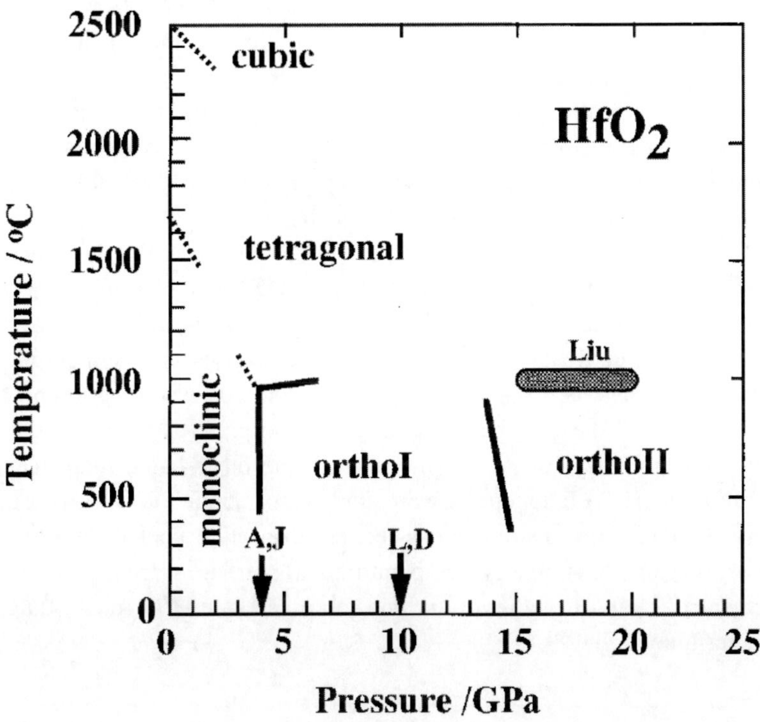

Figure 2. P–T phase diagram for HfO$_2$, showing Monoclinic – orthoI and orthoI – tetragonal and cubic phases. Dotted lines are assumed boundaries. Arrows with letters are monoclinic-to-ortho I transition pressures observed at room temperature: A = Arashi [23], J = Jayaraman et al. [24], L = Leger et al. [25], D = Desgreniers [26]. Reprinted from reference [27] copyright license 3404760760165.

In fact, the 4-fold coordinated O atom in the monoclinic structure is centered within a distorted tetrahedron, whereas the 4-fold coordinated O atom in the cubic structure is at the center of a regularly shaped tetrahedron.

Above 1700°C, similarly to ZrO$_2$, HfO$_2$ becomes progressively tetragonal. This tetragonal structure (P4$_2$nmc) can be regarded as a distorted fluorite-type cubic structure [15, 17]. From 2600°C the tetragonal to cubic transformation occurs.[18] The cubic HfO$_2$ polymorph (Fm3m) has a fluorite type structure in which the oxygen atoms are coordinated to four Hf atoms, each oxygen being at the center of a regularly shaped tetrahedron (Figure 1a).

A forth polymorph of HfO$_2$ can be stabilized under the application of hydrostatic pressure. Figure 2 from ref [20] shows the phase transition from the tetragonal structure to the orthorhombic structure (Pbcm for orthorhombic

I and Pnma for orthorhombic II) at high temperature and under hydrostatic pressure. The stabilization of the orthorhombic polymorph in the bulk materials appears to be difficult without using dopants [20]. However, the orthorhombic structure has been observed in the thin film form and its stabilization was attributed to interface strain and more particularly to the substrate induced strain via the difference of expansion coefficient between the substrate and the film (see part 2) [21]. The various crystal structures of HfO_2 as a function of temperature and pressure were recently calculated by Caravaca et al., using density functional theory (DFT) calculation [22].

In the bulk form, multiple solid solutions of HfO_2 were studied. More particularly, the stabilization of higher symmetry phases via the incorporation of another element was investigated during 60s and 70s. The main problem for the stabilization of these higher symmetry polymorphs was related to the high temperature necessary for their synthesis. On the other hand, more recently, during the 90s, HfO_2 compounds were studied for applications in the nuclear industry for their high neutron cross-section absorption coefficient [14]. The effect of oxygen stoichiometry, temperature and pressure were investigated by Degrenier et al., and Lowther et al., and more accurate HfO_x phase diagrams were determined [26, 28].

HfO$_2$ Based System in the Bulk Form

Divalent elements like Mg, Ca, Ba, Sr were studied. The addition of MgO into the structure was found to stabilize the cubic and tetragonal structures and lower the temperature of transitions [29]. The addition of CaO at a concentration higher than 10mol% induces the formation of a fluorite-like cubic solid solutions $(CaO)_x(HfO_2)_{1-x}$ and also the formation of the $CaHfO_3$ with perovskite-type structure [30]. The addition of CaO also induces the formation of three different phases: $CaHf_4O_9$, $Ca_2Hf_7O_6$, $Ca_6Hf_9O_{44}$ whose structures are derivatives of the face centered cubic structure and differ only by the type of ordering [31]. It was observed that the addition of an alkaline rare earth element in the HfO_2 structure induces the stabilization of the $MHfO_3$ perovskite-type structure (M=Ca, Sr, Ba). However, in the bulk form $CaHfO_3$ and $SrHfO_3$ (Figure 3) both formed orthorhombically distorted perovskite structures (Pnma) [32], whereas $BaHfO_3$ was found to possess the cubic perovskite structure (P$m3m$) [33]. Recent studies demonstrated that the synthesis of the perovskite in the nanoparticle form allows the stabilization of the cubic (P$m3m$) structure of $CaHfO_3$ and $SrHfO_3$ (Figure 3, red stars) [34].

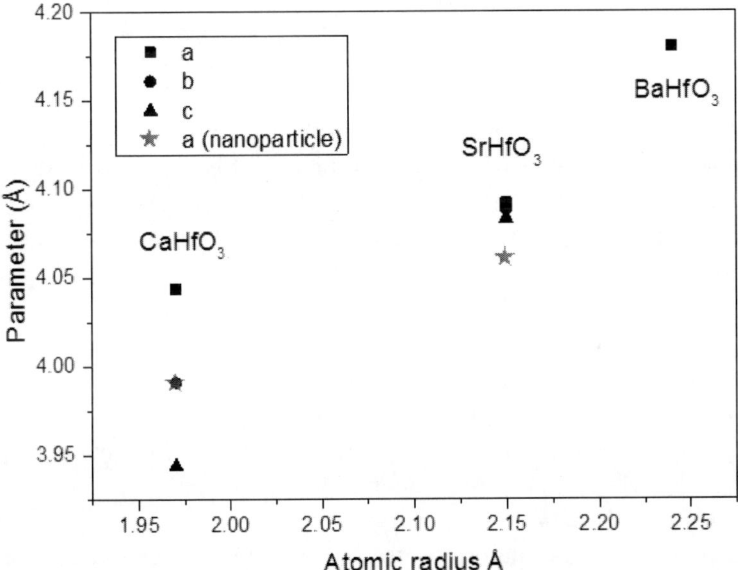

Figure 3. The lattice constants a (black squares), b (black circles) and c (black triangles) of the primitive bulk cell for CaHfO$_3$, SrHfO$_3$ with Pnma orthorhombic structure and BaHfO$_3$ with P$m3m$ cubic structure at RT from [32, 33] and lattice constants a (red stars) of the nanoparticles cells for cell for CaHfO$_3$, SrHfO$_3$ with P$m3m$ cubic structure from [34].

As in the case of HfO$_2$ compound, the phase transition of ZrO$_2$ induces large volume changes and hinders applications at higher temperatures. So, a lot of efforts were involved in the stabilization of the high temperature cubic structure at ambient conditions via the incorporation of appropriate metal oxide additives and avoiding the non-desirable cubic to tetragonal to monoclinic phase transformation. For this reason, yttrium stabilized zirconia (ZrO$_2$-Y$_2$O$_3$) was one of the most studied systems. In the same way, HfO$_2$-Y$_2$O$_3$ was intensively studied [35]. The addition of Y$_2$O$_3$ induces the stabilization of the cubic phase. In fact, as already discussed, the cubic HfO$_2$ structure is a fluorite-type structure [1] and Y$_2$O$_3$ crystallizes in a cubic-centered structure type. These two structures are closely related [36] and the phase diagram of HfO$_2$–Y$_2$O$_3$ [1, 37] shows that, depending on the preparation conditions, it is indeed possible to stabilize HfO$_2$ in the fluorite structure. Caillet et al., showed that it is possible to stabilize the cubic phase without any presence of the monoclinic phase from 8 mol.% of Y$_2$O$_3$ at 1800°C [38]. The stabilization of the cubic phase was also studied by Schieltz et al., who reported a phase boundary between 6 and 8 mol.% Y_2O_3 at 1500°C [39]. HfO_2 cubic solid solution was reported to be stable under a wide compositional range from 8/15% up to 50/60% for temperatures of 1500°C or higher [37, 38].

The addition of lanthanide rare-earth elements like Erbium (Er), Lanthanum (La), Cerium (Ce) and Ytterbium (Yb), was also studied. The phase diagrams are rather complex with multiple phases and structures being simultaneously present. The addition of lanthanide elements, allows the formation of higher symmetry polymorphs, but not without the simultaneous formation of other phases [40].

Of similar interest, TiO_2 compounds have many different fields of application and the study of the addition of TiO_2 in HfO_2 was of high interest. The solubility of TiO_2 in HfO_2 is rather high, more specifically in the higher symmetry polymorphs. The phase diagram is relatively complex with the presence of eutectic, metatectic, eutectoid points and an incongruent melting point of $HfTiO_4$ phase [41].

Many other solid solutions were also studied. The fabrication of new toughened ceramics and more particularly hafnia based toughened ceramic was the main goal of these first studies on the different hafnia solids solutions. The high melting point of the HfO_2 compound (700°C higher than ZrO_2 compound) was one of the reasons for these efforts put into the study of these solid solutions.

HfO$_2$ THIN FILMS

Pure HfO$_2$ Thin Films

The investigations on superior materials for the replacement of SiO_2 as a gate dielectric material in CMOS technology were focused on insulating materials compatible with silicon presenting a high stability against crystallization, the lack of which is in fact the main reason for the generation of leakage currents. The latter in fact is triggered due to the formation of grain boundaries on grain growth and crystallization. Moreover, the replacement material should not present any undesirable phase transitions during their processing or integration into devices. For these reasons, HfO_2 has received significant attention and a regain of interest since the beginning of 2000 [2a, 2c, 42]. In addition, compounds like HfO_2 presenting higher permittivity values, were also of interest for the fabrication of highly integrated metal-insulator-metal (MIM) capacitors with high-capacitance values. HfO$_2$ was more particularly studied because of its higher thermal stability on Silicon substrates compared to ZrO$_2$ and more specifically, compared to lanthanide compounds and Y$_2$O$_3$ [43]. This thermal stability is important, more specifically when the films are subjected to annealing temperatures under 1000°C in O$_2$ or NH$_3$ and do not show any evidence of forming natural silicate compounds at these temperatures [12]. HfO$_2$ was then intensively studied during the last 10 years, in the thin film form; a lot of efforts were put into finding a method for the stabilization of higher symmetry phases that present higher dielectric permittivity. In fact, *ab initio* calculations demonstrated that higher-symmetry polymorphs of HfO$_2$, exhibit higher permittivities than the thermodynamically stable, room temperature monoclinic phase. Zhao and Vanderbilt have theoretically predicted that the cubic and tetragonal HfO$_2$ structures should exhibit a higher permittivity of 29 and 70 respectively, compared to the permittivity of the monoclinic structures 16–18 [44]. More recently, the DFT calculations published by Rignanese showed similar results. It is therefore desirable to synthesize HfO$_2$ in the cubic, tetragonal or orthorhombic forms for CMOS technology and MIM capacitor applications. However, as we saw in the first part of this chapter, these higher symmetry polymorphs are only stable at high temperatures and higher pressures which are in fact, necessary conditions for the stabilization of the orthorhombic structure [20, 27]. Furthermore, monoclinic to tetragonal and tetragonal to cubic transformations occur at, respectively, 1700 and 2700°C.

Like for ceramics, HfO$_2$ thin films exhibit multiple crystallographic structures; contrary to ceramics, the orthorhombic structure can be stabilized in the thin film form without the application of high pressure. The monoclinic structure is the main crystalline phase observed in thin films at room temperature [45]. However, cubic and tetragonal polymorphs have been observed simultaneously with the stable monoclinic structure [46]. The oxide thin films are known for their natural oxygen understoichiometry, also a prerequisite for the stabilization of higher symmetry phases. However, in the thin film form, the interface strain and difference of expansion coefficients should also be taken into consideration. The stabilization of the higher symmetry polymorphs can be brought about by two phenomena. First, a difference of expansion coefficient can induce a compressive strain which will in turn stabilize the lower unit cell volume of the tetragonal, orthorhombic or cubic phases. In fact, the thermal expansion coefficient of HfO$_2$ (5.8.10^{-6}K^{-1})^{15} is 10 times higher than the coefficient of SiO$_2$ (0.5.10^{-6}K^{-1}). So, the combination of the ferroelastic nature of transitions between the HfO$_2$

polymorphs and the thermal expansion mismatch that place the film under biaxial tension will in fact, induce compressive strains on the film that will stabilize the lower unit cell volume or a higher symmetry polymorph. This can explain the stabilization of the orthorhombic structure in the case of the thinnest films [21]. The second phenomenon is an oxygen sub-stoichiometry in the thinnest films. In fact, the existence of oxygen vacancies in HfO_2 thin films could lead to the stabilization of the tetragonal or orthorhombic structures [47]. Manory et al. showed that oxygen vacancy is a key factor for the stabilization of the orthorhombic or cubic structures [48]. HfO_2 orthorhombic polymorphs were then observed in thin films deposited at 500°C using atomic layer deposition (ALD) [49]. Cubic nanocrystallites were also observed inside HfO_2 monoclinic thin films by the same group [50]. The influence of the interface strain was illustrated by Shandalov et al., where they showed the phase transitions from amorphous to tetragonal HfO_2 and further on to the monoclinic HfO_2 with the increase in the thickness of the coating [51]. Many groups have already observed that generally the crystallization starts with a tetragonal or orthorhombic structure. This is followed by the observation of a mixed phase monoclinic structure and finally by the progressive phase change into a single-phase monoclinic structure with the increase of the film thickness or of the growth temperature [52]. Similar mixed phases have also been observed for a growth temperature of 500°C using other precursors [53]. Aarik et al., also reported that at the beginning of the growth, the formation of a mixed orthorhombic, tetragonal, cubic and monoclinic structure is usually observed, and then by increasing the thickness the more stable structure (i.e., monoclinic polymorph) is promoted. They showed the appearance of the cubic structure for films thinner than 10nm grown at high temperatures [50]. This gives an idea of how difficult it is to predict the phase that will be stabilized in the case of thin films, as it depends on the growth conditions (temperature, pressure, gas nature…), the method of deposition, the thickness of the film and the nature of the substrate.

The growth at low temperatures usually favours a metastable amorphous phase in which nanocrystallites can be embedded as discussed before. Generally a slightly lower roughness is then observed at lower deposition temperatures <475°C, which might be related to the lack of crystalline grain growth at these temperatures. Aarik et al., also suggested that dimensions of the crystallites in HfO_2 films are limited by film thickness and the appearance of metastable polymorphs in the thin films indicate that these phases may be stabilized because of size effects and/or intrinsic strains [54].

Physical Vapor Deposition

Pulsed Laser Deposition (PLD) method: This very flexible method that allows a direct stoichiometric transfer from the target to the deposited thin films is however more adapted for fundamental research studies due to the limited surface area that can be coated via this technique.

The HfO$_2$ thin films deposited using the PLD method are usually of polycrystalline nature [55]. The films can also be amorphous or monoclinic if deposited at temperatures lower than 500°C [8b]. However, the transition of the amorphous films into an orthorhombic phase upon annealing under N$_2$ was also observed [56].

Radio-Frequency sputtering: He et al., illustrated that HfO$_2$ thin films deposited via RF-sputtering are amorphous. Nevertheless, HfO$_2$ films transform into tetragonal and then into monoclinic phases upon annealing with the increase of the annealing temperature to finally form a polycrystalline structure at high annealing temperatures (Figure 4a) [57].

The HfO$_2$ thin films were mainly amorphous or monoclinic and depend on the gas mixture (Ar/O$_2$) [58]. More recently, Migita et al., showed that in the case of films deposited at room temperature via RF-sputtering, the stabilization of the higher symmetry phase also depends on the starting phase. They observed that under annealing, amorphous HfO$_2$ transforms into cubic HfO$_2$ polymorph, but as soon as the monoclinic phase nucleates, cubic transformation no longer becomes possible [59].

Similar results were obtained by Li et al., and they introduced an unreported cubic like short range order in the amorphous HfO$_2$ structure of the HfO$_2$ films deposited by RF-sputtering [60].

The study of the crystallographic structure of the films revealed the presence of a cubic short range order in the amorphous structure of the film. In the amorphous phase, the Hf atoms are eightfold coordinated with fourfold coordinated O similarly to the cubic structure and not sevenfold coordinated for Hf; they are three and fourfold coordinated for O similarly in the monoclinic structure.

This short range order serves as a nucleation site for topotactic rearrangement [61] of the atoms into a cubic structure upon high temperature annealing. Rauwel et al., observed that amorphous HfO$_2$ thin films deposited on Mg interlayer transforms into cubic polymorph upon N$_2$ annealing [62] due to the diffusion of the Mg atoms from the interlayer between the HfO$_2$ film and the silicon substrate via a Kirkendall type effect (Figure 4b) [63].

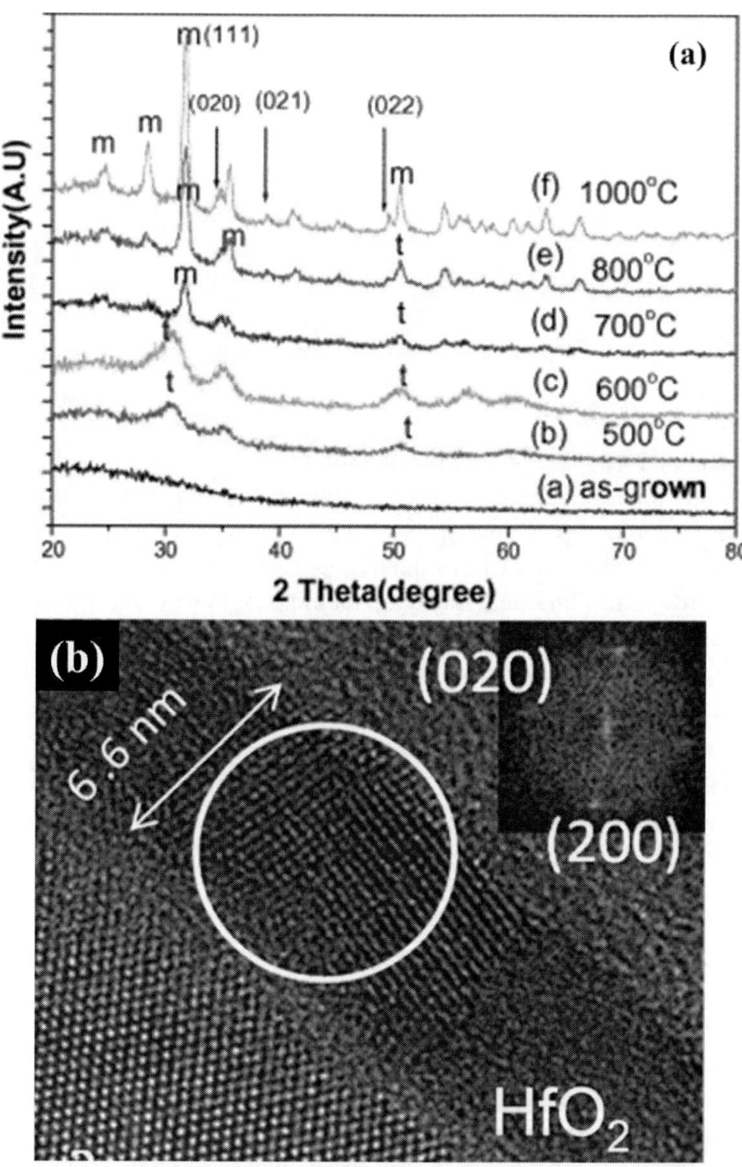

Figure 4. (a) XRD pattern of the as-grown and annealed HfO$_2$ films at various temperatures for 3 min in O$_2$ ambient. The letters m and t represent the monoclinic and tetragonal structure of the HfO$_2$ film, respectively. Reprinted from [57] copyright license 3412961201662, (b) HRTEM image of annealed HfO$_2$ thin film with RTA treatment at 800°C for 1 min under N$_2$. The inset is a fast Fourier transform of the encircled grain presenting the HfO$_2$ cubic phase and oriented along the b001> zone axis. Reprinted form reference [62b] copyright license 3412961318161.

Chemical Vapor Deposition

During the last decade, in the numerous publications reported on the deposition of HfO$_2$ for microelectronic applications, chemical vapor deposition process (CVD) and atomic layer deposition (ALD,) have been widely studied and used for the growth of the films. Metal organic chemical vapor deposition (MOCVD) allows the growth of a film over a large surface area with a good composition control and uniformity which offer the advantage of possible deposition at the industrial scale on large wafers with conformal step coverage. More particularly, ALD was demonstrated as the most promising route for conformal coating of complex nanostructures and nanoporous structures without blocking the nanoporosity [64]. ALD is a unique thin film deposition technique based on alternate surface-controlled reactions from the gas phase, which provides conformal precise thickness control and the self-limiting deposition process. Chemical vapor deposition methods also allow for good composition and thickness control of uniform films when appropriate chemical precursors are used.

Chemical vapor deposition: Most of the studies on the deposition of HfO$_2$ based on the decomposition of metalorganic chemical precursor (halides, alkoxides, amides) showed that when grown at low temperatures, HfO$_2$ is amorphous [65] and is monoclinic when grown at higher temperatures[45b, 66]. However, some investigations also demonstrated that kinetic factors and nucleation can lead to the formation of higher symmetry polymorph under certain conditions [67]. Schaeffer et al., showed that for the same precursor, the crystalline structure of the HfO$_2$ thin films can vary with the deposition temperature. Like previously discussed, the phase stabilized during the deposition process will lead to the formation of a crystalline phase upon annealing. They observed that annealing treatment induces the formation of the tetragonal structure if the deposited film is amorphous or tetragonal; whereas for the monoclinic phase present during growth, the annealing treatment only enhanced the crystallinity of the films [68]. Many precursors were used and studied [45b, 69], Teren et al., compared alkoxide and amides precursors, but the structure of the films were quite similar; amorphous for low deposition temperature and monoclinic for temperature higher than 450°C, but observed differences in the electrical properties due to the nature of the precursor [70]. Liquid injection MOCVD was presented as an innovative CVD technique [71] that can avoid the usage of poorly volatile precursors and offer a better control of the amount of liquid precursor injected during the deposition process which also determines the thickness of the deposited thin film [45b]. It was also demonstrated that the injection frequency of the precursor has an important influence on the polymorph that will be stabilized during the deposition process and by the appropriate choice of injection frequency it is possible to stabilize the appropriate higher symmetry polymorph [72].

Atomic Layer Deposition: ALD is now a leading process for the deposition of ultra-thin films with a high conformality. This deposition is even more studied due to its unique ability of coating architecture in three dimensions. Many different ALD processes are now available; Conventional ALD, direct plasma ALD, remote plasma ALD, enhanced plasma ALD. It was observed that the use of plasma enables a better decomposition of the precursor and promotes the crystallization of the film in the monoclinic structure [73]. ALD of HfO_2 thin films is in fact certainly one of the most studied fields among the other high-k materials or other deposition methods studied in that field. The literature is very rich and it is not possible to describe all the work performed on that subject. As already observed in the case of MOCVD, most of the films deposited at low temperature via ALD are amorphous [13, 74]. Depending on the precursor used for the deposition, monoclinic structure will appear as the predominant crystalline phase [50, 75]. However, like in the case of thin films deposited via MOCVD, higher symmetry polymorphs can be observed in the case of specific growth conditions, in very thin films and via annealing processes of amorphous HfO_2 thin films [76]. In order to improve the quality of the films, a thermal treatment is usually required after the ALD cycles to enhance the crystallinity or reduce the number of intrinsic defects related to the process of deposition itself [74b, 74c, 77]. More recently, Cho et al., investigated the influence of oxygen in ALD using a low active oxygen source that promotes the formation of carbonate and stabilizes the HfO_2 tetragonal polymorph [47].

Sol-Gel Method

In the case of sol-gel methods, the HfO_2 thin films are amorphous. It was reported that the HfO_2 thin films only crystallize in the monoclinic structure at temperatures higher than 500°C or upon annealing [78]. No cubic like, short range order was detected in the amorphous HfO_2 structure of the films deposited by sol-gel method. This is maybe due to the deposition method itself, using metalorganic precursors like halides ($HfCl_4$) or alkoxides.

Despite all these available deposition methods and process the reliable stabilization of the higher symmetry polymorph remains challenging in the case of pure HfO$_2$ thin films. In fact, many parameters have a simultaneous influence during the growth of the film and it is really difficult to predict the phase that will be stabilized as it depends on the deposition method, the growth conditions (temperature of deposition, gas pressure, gas mixture nature…), thickness of the film and the nature of the substrate on which the film will be deposited. For these reasons, the stabilization of the higher symmetry polymorphs by the addition of another metal oxide into the HfO$_2$ structure has initiated many investigations in the high-k material community.

Doped HfO$_2$ Thin Films and Stabilization of Higher Symmetry Polymorphs

The cubic structure of HfO$_2$ can be stabilized by the addition of another metal oxide element with an oxidation state lower than 4+ via the formation of a solid solution. In fact, the charge compensation for added cations of lower charge is achieved by the creation of a number of vacant sites on the anion sublattice that depends on the oxidation states of the added element. One of the advantages of zirconia and hafnia structure is the important tolerance to the presence of oxygen vacancies inside the structure. For these reasons, the influence of the addition of a metal oxide in HfO$_2$ thin film such as MgO, Al$_2$O$_3$, SiO$_2$, CeO$_2$, Sc$_2$O$_3$, Y$_2$O$_3$ and most of the rare-earth oxides have been studied. Depending on the metal oxide that will be added into the HfO$_2$ structure, easier stabilization of higher symmetry structure at lower temperature may be observed. However, for certain metal oxides, a delay of crystallization can also be observed, keeping HfO$_2$ amorphous up to 900°C can be desirable in the case of oxide dielectric gate applications thus allowing us to achieve low leakage current densities. The addition of metal oxide for delaying the crystallization of HfO$_2$ will not be developed in this chapter, but some references will be provided.

The addition of Y$_2$O$_3$ into HfO$_2$ thin films spurred a lot of interest. In fact, in the case of Y$_2$O$_3$ addition, the position of the oxygen vacancies in the structure will be different compared to other cations. Usually, the Coulomb interaction between the charged defects encourages oxygen to become the nearest neighbor, but the bigger size of Y compared to Hf and Zr makes the second nearest neighbor more favorable [79]. This important particularity makes the oxygen vacancies the nearest neighbor of the Zr and Hf atoms, which induces the stabilization of the higher symmetry polymorph and more particularly the stabilization of the cubic polymorph as we already discussed in the previous part. For thin films, many studies have been devoted to the HfO_2–Y_2O_3 system, but only limited selected compositions were investigated [80]. Rauwel et al. described a complete study of Y_2O_3 addition into HfO_2 thin films deposited on silicon by metal organic chemical vapor deposition over a wide composition range [80a, 81]. They demonstrated at first the simultaneous presence of the monoclinic structure mixed with a higher symmetry phase. The stabilization of the pure cubic phase was observed from 6.5at.%. The variation of the structural properties with the addition of yttrium oxide was described in detail and the formation of yttrium silicate was observed for an yttrium addition of 12.5at.%; the amount of silicate and Y_2O_3 compounds becoming more and more prominent with the continuous increase of Y_2O_3 content. Wang et al., compared the effect of the incorporation of 10at% Y and Al into HfO_2 structure. They observed a monoclinic structure for the pure HfO_2 films, only cubic phase existed in the Y-incorporated HfO_2, whereas Al-incorporated HfO_2 film appears to be of amorphous nature [80i]. They conclude that both types of doping introduce significant amount of oxygen vacancies along with different results in the structural and electrical properties. Like for Y_2O_3, the addition of Sc_2O_3 was studied. Yakovkina et al., demonstrated that similar to Y_2O_3, depending on the scandium content in the films, the structure changes from monoclinic to cubic [82]. Brizé et al., reported the stabilization of the cubic polymorph with 10at.% and under forming gas annealing at 520°C [83].

The studies of dopant element with a smaller ionic radius than Hf element such as SiO_2, GeO, SnO_2 or TiO_2, show that the addition of these metal oxide compounds into the HfO_2 structure, promotes the stabilization of the tetragonal phase. However, this addition should be of a moderate level ~8-12at.%. It was observed that the increase in the concentration of these metal oxide dopants, usually induces the stabilization of the tetragonal phase but also leads to phase separation if the concentration is too high [84].

For other metal oxides added into the HfO_2 structure, oxygen vacancies are the nearest neighbors of the added element. However, the charge compensation for added divalent cations will be more important than trivalent cations and the creation of a number of vacant sites in the case of divalent cations will be twice higher. Two trivalent cations are necessary for the formation of one oxygen vacancies. This makes the addition of divalent cations normally more efficient for the stabilization of a higher symmetry HfO_2 polymorph.

The addition of lanthanides was also studied. Ushakov et al. reported that the addition of lanthanum oxide into the HfO$_2$ structure induces the stabilization of the cubic structure, but on further heating, the solid solutions decompose into the pyrochlore structure (Hf$_2$La$_2$O$_7$) and monoclinic HfO$_2$ [85]. Crystallization delay and formation of pyrochlore phase were also reported by Toriumi et al. [86]. Yamamoto et *al.*, also observed the stabilization of the cubic fluorite structure of Y$_2$O$_3$-doped HfO$_2$ via the addition of La$_2$O$_3$ into HfO$_2$ structure [87]. In a similar manner, Dy, Er and Gd addition were studied by Adelmann et al., Govindarajan et al., and Losovyj et al., respectively [88]. They observed that the tetragonal polymorph can be stabilized with 10at.% Er or 10at.% Dy under high temperature annealing; whereas, in the case of 10at.% of Gd addition, the cubic structure is stabilized, under similar annealing conditions. The stabilization of the cubic phase also was observed for 10at.% Dy addition, but under annealed in forming gas at 520°C.

MgO addition, has not been intensively studied, Rauwel et al. observed that the monoclinic structure is dominant until a Mg content of 16.6at.%. Pure cubic structure is then observed for Mg content of 22.3at.%, but for higher Mg content $\geq$ 30at.% XRD indicates that films are amorphous. However, the presence of nanocrystallites inside the amorphous cannot be neglected [89].

The addition of magnetic cations for the stabilization of cubic HfO$_2$ thin film was not extensively studied. The main reason was that the targeted application as gate dielectric oxide was not achievable. Some non-magnetic transition elements were however studied. Lu et al., report that Ta and Al additions have similar effect [90]. The Ta-Hf-O films are amorphous and Ta addition improves the thermal stability of the films [91]. The films moreover remain amorphous after high temperature annealing (950°C for 30s) [92].

The stabilization of higher symmetry HfO$_2$ polymorphs in the thin form is still under investigation and is of importance in the field of microelectronic applications. The stabilization of pure cubic HfO$_2$, does not appear easy and to our knowledge and hence, no reliable industrial method has been found to date. This is because; the stabilization of higher symmetry phases depends on the combination of multiple parameters (method, growth conditions, oxygen stoichiometry, thickness...). The addition of metal oxide compounds into the HfO$_2$ structure was regarded as the most promising solution. However, the introduction of another element in the structure also affects the intrinsic properties of the films. These solutions are more like a compromise; the modification of the properties via the addition of another element into the HfO$_2$ structure has to be taken into consideration. Some elements like Mg will simultaneously stabilize the cubic structure and have some beneficial effects on the electrical properties, but other elements like Y will modify the dielectric properties and also promote the formation of silicates into the structure.

HfO$_2$ NANOPOWDER

Pure HfO$_2$ Nanoparticles

Filipovich et al., demonstrated that it is possible to stabilize a high temperature polymorph of a crystal below its normal transformation temperature at some critical crystallite size [93]. This observation then demonstrates the possibility of stabilizing higher symmetry polymorphs in the nanosize. The first HfO$_2$ nanopowder was synthesized by Mazdiyasni et al., using hafnium alkoxides precursors via either vapor-phase decomposition or hydrolytic decomposition [94]. High purity HfO$_2$ nanoparticles with size ranging from 5-20nm size were prepared.

Many studies were performed on the preparation of pure HfO$_2$ using different synthesis routes. All synthesized HfO$_2$ nanoparticles are crystalline due to their high surface energy. However, higher symmetry polymorphs of pure HfO$_2$ were only reported recently. Diverse routes of synthesis were investigated, but all nanoparticles were monoclinic. Monoclinic HfO$_2$ nanorods were synthesized by Tirosh et al., via non-hydrolytic sol-gel method [95]. Hydrothermal reaction-sintering also induces the formation of monoclinic HfO$_2$ nanoparticles [96]. More recently, monoclinic rice-like HfO$_2$ nanostructures were obtained by the microwave-hydrothermal method, using halide precursors [97]. Stefanic et al., studied the synthesis of HfO$_2$ and ZrO$_2$ nanoparticles using conventional hydrothermal methods. They observed that in the case of ZrO$_2$, it is possible to obtain higher symmetry polymorphs and they synthesized the metastable tetragonal and cubic phases of ZrO$_2$ in the nanoparticle form. However, they also observed a different chemical mechanism and no higher symmetry polymorph could be stabilized via these methods. They concluded that the topotactic model of crystallization proposed for ZrO$_2$ cannot be used in the case of HfO$_2$ [98]. A similar result was observed using non-aqueous sol-gel based on benzyl alcohol route and only the monoclinic structure was observed for the synthesized HfO$_2$ nanoparticles [99]. However, in a recent study, Rauwel et al., demonstrated that by using adapted precursors, it is possible to stabilize separately 2 different polymorphs of HfO$_2$: monoclinic and cubic (Figure 5) [19]. In fact, the cubic phase cannot be stabilized without oxygen vacancies and by using a reductive solvent such as benzylamine with Hf terbutoxide precursor, the stabilization of the cubic phase was possible. Moreover, both the monoclinic and the cubic samples were pure without any indication of carbonates. The intrinsic defects and oxygen vacancies formed during the synthesis induces photoluminescence response upon UV exposure [100] confirming that the introduction of oxygen vacancies into the structure promotes the stabilization of higher symmetry polymorph in the case of pure cubic HfO$_2$ nanoparticles.

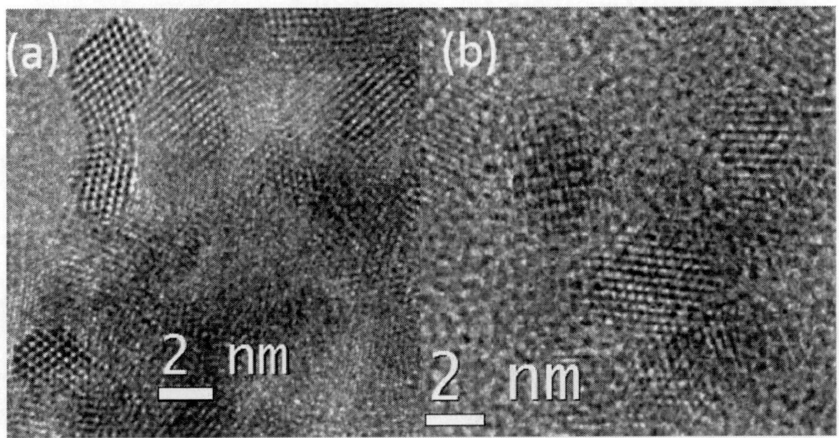

Figure 5. HRTEM image of HfO$_2$ (a) cubic and (b) monoclinic nanoparticles.

Doped HfO$_2$ Nanoparticles

Like in the case of thin film deposition, the addition of other metal oxides into the HfO$_2$ structure may promote the stabilization of higher symmetry phases of HfO$_2$ nanoparticles. Contrary to thin films deposition and applications, the goal of the addition of another element during nanoparticle synthesis is not specifically for the synthesis of a higher symmetry phase. The addition of the other elements aims at adding new properties like photoluminescence for scintillators or magnetism for diluted magnetic semiconductor (DMS). However, the cubic phase is also desirable in the case of optical applications as the cubic symmetry is mandatory due to its isotropic optical properties which avoid scattering due to abrupt refractive index changes at grain interfaces.

Similarly to pure HfO_2 nanoparticles, Mazdiyasni et al., prepared the first Y_2O_3-stabilized HfO_2 powders using alkoxide precursors [101]. The introduction of transition metal in order to add magnetic properties into the HfO_2 nanoparticles also affected their structural properties. The addition of 10at.% of Mn is sufficient for the stabilization of the cubic structure. In the case of Cr addition, it was observed that 7at.% is not sufficient for the appearance of the cubic polymorph. However, the authors did not cover a wide range of doping and it is then difficult to conclude on the effect of the quantity of transition metal into the HfO_2 structure [102].

The influence of the addition of several rare-earths into monoclinic HfO_2, prepared by hydrothermal and Pechini-based sol-gel methods was reported by Tanigushi et al., and Wiatrowska et al. [103]. More recently, Lauria et al., studied the effect of the addition of rare earth into the HfO_2 structure for the enhancement of photoluminescence properties. They report that the pure cubic phase can be stabilized from 8at.% of Eu and 10at.% of Lu together [104].

CONCLUSION

This chapter is a compilation of different works aiming at achieving higher symmetry phase stabilization of HfO_2. Since its discovery, HfO_2 has spurred a lot of interest and investigations on its different solid solutions with other metal oxide compounds and on the stabilization of higher symmetry polymorphs, have taken a leap. The effect of oxygen stoichiometry, pressures and temperatures during growth have been studied in bulk HfO_2 right from the 70s. In the present day, HfO_2 is considered as a material of technological importance for the microelectronic industry and more particularly for the replacement of Si-based oxide dielectric gate in transistor technology. Intel has already integrated hafnium oxide in commercial CMOSFETs since 2007. The stabilization of higher symmetry phase is of particular interest as these phases exhibit higher dielectric constants and has incited a lot of curiosity among researchers in the field.

It has been demonstrated that during thin film deposition (CVD, ALD, RF-sputtering), multiple parameters have to be taken into consideration like growth conditions (temperature, oxygen partial pressure, pressure…), the thickness of the film, the size of the crystallites inside the film and the type of substrate on which HfO_2 will be deposited. Post-deposition annealing can also affect the structure of the deposited films depending on the gas environment (forming gas, N_2, O_2, NH_3) and the temperature that will be used. Some reports have suggested that once the monoclinic phase nucleates, crystallization of all other phases of HfO$_2$ are inhibited. These investigations demonstrated that the stabilization of pure HfO$_2$ higher symmetry phase is not easy. The oxygen sub-stochiometry was considered as the key factor for the stabilization of higher symmetry phases. The addition of metal oxides into the HfO$_2$ structure was then considered as the most reliable solution for the stabilization of the cubic phase and many efforts were put into this made to achieve this goal. However, the addition of other elements will also affect the intrinsic properties of HfO$_2$ thin films and compromises have to be made.

Contrary to thin films, in the nanocrystalline form, the stabilization of the cubic phases was not considered of importance except in the case of optoelectronic applications, where a cubic symmetry is desirable due to its isotropic optical properties.

Recent reports show that using reductive precursors does incite the stabilization of higher symmetry phases via the formation of oxygen vacancies and these HfO$_2$ nanoparticles exhibit interesting optoelectronic properties due to the presence of point defect states within the band gap. The main goal of doping the HfO$_2$ nanoparticles was to obtain new properties (i.e., photoluminescence or magnetism) that are not specifically related to microelectronic applications. However, if the addition of other elements can stabilize higher symmetry phases, secondary phases are not uncommon in such cases and therefore cannot guarantee a monophasic material.

PARTICIPANTS

Protima Rauwel obtained her PhD in 2005 from the University of Caen, France in Condensed Matter Physics during which she specialized in optoelectronic materials. She then continued in 2006, with a Post-Doc at the University of Aveiro, Portugal where she studied thin films including HfO$_2$ and nanoparticles via Transmission Electron Microscopy for various applications.

In 2009 she moved over to the University of Oslo, Norway, as a Researcher, where she continued her activity with functional oxides and HfO$_2$. In 2013, she accepted a position at the University of Tartu, Estonia as a founding Researcher of Scanning Transmission Electron Microscopy and also continues working with Optoelectronic materials and their applications, such as HfO$_2$.

Erwan Rauwel, obtained his PhD of Materials Chemistry in 2003 and his Habilitation in 2013 from the University of Caen, France). In 2004, he moved to the LMGP-INP (Minatec), Grenoble, France and worked closely with STMicroelectronics on the deposition of HfO_2 thin films. In 2006, he received a Marie Curie Fellowship (IEF) and joined Department of Chemistry and CICECO of the University of Aveiro, Portugal where he invented and developed a new method of deposition for HfO_2 via ALD. In 2009, he moved to the University of Oslo in the Department of Chemistry (Norway) as senior researcher and worked closely with Statoil Company. Since 2012, he has been appointed Professor at Tartu College, Tallinn University of Technology (TTÜ) (Tartu, Estonia) with a chair of "Sustainable Energy". His research activity is focused on the thin film deposition and the synthesis of nanomaterials using sol–gel and their direct applications.

ACKNOWLEDGMENTS

The authors acknowledge Dr. F. Ducroquet from LETI-CEA, Grenoble for fruitful discussion. ER would like to thank the Estonian Ministry of Education and Research (Project PUT-431) and Doctoral Studies and Internationalization Program 'DoRa' Activity 2 'Improving the quality of higher education by supporting the employment of international teaching staff' receiving funding from the European Social Fund under project 1.2.0201.08- 0001 in Estonia.

REFERENCES

[1]   Wang, J.; Li, H. P.; Stevens, R., Hafnia and hafnia-toughened ceramics. *J Mater Sci* 1992, 27 (20), 5397-5430.

[2]   (a) Robertson, J., High dielectric constant oxides European Physics Journal of Apllied Physique 2004, 28, 265; (b) Wallace, R. W.; Wilk, G., High-k Dielectric Materials. *MRS Bulletin* 2002, 3, 192; (c) Robertson, J., High dielectric constant gate oxides for metal oxide Si transistors. *Reports on Progress in Physics* 2006, 69 (2), 327-396.

[3]   Fiorentini, V.; Gulleri, G., Theoretical Evaluation of Zirconia and Hafnia as Gate Oxides for Si Microelectronics. *Physical review letters* 2002, 89 (26), 266101.

[4]   (a) Sayan, S.; Aravamudhan, S.; W. Bush, B.; Schulte, W. H.; Cosandey, F.; Wilk, G. D.; Gustafsson, T.; Garfunkel, E., A study of HfO_2 film interfaces with Si and SiO2. *Journal of Vacuum Science & Technology, A: Vacuum, Surfaces, and Films* 2002, 20 (2), 507; (b) Kuo, C. T.; Kwor, R.; Jones, K. M., Study of sputtered HfO_2 thin films on silicon. Thin Solid Films 1992, 213 (2), 257; (c) Hu, H.; Zhu, C.; Lu, Y. F.; Li, M. F.; Cho, B. J.; K., C. W., A high performance MIM capacitor using HfO2 dielectrics. *IEEE Electron Device Letters* 2002, 23, 514; (d) Kang, L.; Lee, B. H.; Qi, W.-J.; Jeon, Y.; Nieh, R.; Gopalan, S.; K., O.; Lee, J. C., High quality Ultra thin CVD HfO_2 Gate stack with Poly-Si Gate electrode. *IEEE Electron device letters* 2000, 21 (4), 181.

[5]   Wilk, G. D.; Wallace, R. M.; Anthony, J. M., High-kappa gate dielectrics: Current status and materials properties considerations. *Journal of Applied Physics* 2001, 89 (10), 5243.

[6]   (a) Gutowski, M.; Jaffe, J. E.; Liu, C.-L.; Stoker, M.; Hegde, R. I.; Rai, R. S.; Tobin, P. J., Thermodynamic stability of high-K dielectric metal oxides ZrO_2 and HfO_2 in contact with Si and SiO_2. *Applied Physics Letters* 2002, 80 (11), 1897-1899; (b) Kingon, A. I.; Maria, J.-P.; Streiffer, S. K., Alternative dielectrics to silicon dioxide for memory and logic devices. *Nature* (London) 2000, 406, 1032.

[7]   (a) Kauerauf, T.; Govoreanu, B.; Degraeve, R.; Groeseneken, G.; Maes, H., Scaling CMOS: Finding the gate stack with the lowest leakage current. *Solid-State Electronics* 2005, 49 (5), 695; (b) Cheng, X.; Song, Z.; Jiang, J.; Yu, Y., Characteristics of sandwich-structured $Al_2O_3/HfO_2/Al_2O_3$ gate dielectric films on ultra-thin silicon-on-insulator substrates. *Applied Surface Science* 2005, 252 (5), 1876.

[8]   (a) Ding, S.-J.; Zhu, C.; Li, M.-F.; Zhang, D. W., Atomic-layer-deposited $Al_2O_3/HfO_2/Al_2O_3$ dielectrics for metal-insulator-metal capacitor applications. *Applied Physics Letters* 2005, 87 (5), 053501; (b) Hu, H.; Zhu, C.; Lu, Y. F.; Wu, Y. H.; Liew, T.; Li, M. F.; Cho, B. J.; Choi, W. K.; Yakovlev, N., Physical and electrical characterization of HfO_2 metal--insulator--metal capacitors for Si analog circuit applications. *Journal of Applied Physics* 2003, 94 (1), 551; (c) Kim, W.-K.; Kang, S.-W.; Rhee, S.-W., Atomic layer deposition of zirconium silicate films using zirconium tetra-tert-butoxide and silicon tetrachloride. *Journal of Vacuum Science & Technology A: Vacuum, Surfaces, and Films* 2003, 21 (5), L16-L18.

[9]   (a) Ding, S.-J.; Zhang, M.; Chen, W.; Zhang, D. W.; Wang, L.-K.; Wang, X. P.; Zhu, C.; Li, M.-F., High density and program-erasable

metal-insulator-silicon capacitor with a dielectric structure of $SiO_2/HfO_2/Al_2O_3$ nanolaminate/Al2O3. *Applied Physics Letters* 2006, 88 (4), 042905; (b) van Schaijk, R.; van Duuren, M.; Yuet Mei, W.; van der Jeugd, K.; Rothschild, A.; Demand, M., Oxide-nitride-oxide layer optimisation for reliable embedded SONOS memories. *Microelectronic Engineering* 2004, 72 (1-4), 395.

[10] Boganov, A. G.; Rudenko, V. S.; Makarov, L. P., X-ray study of zirconium and hafnium dioxides at temperature up to 2750°C. Dokl. Acad. *Nauk.* SSSR 1965, 160, 1065-1068.

[11] Hevesy, G., The Discovery and Properties of Hafnium. *Chemical Reviews* 1925, 2 (1), 1-41.

[12] (a) Wilk, G. D.; Wallace, R. M.; Anthony, J. M., High-κ gate dielectrics: Current status and materials properties considerations. Journal of Applied Physics 2001, 89 (10), 5243-5275; (b) Lin, Y.-S.; Puthenkovilakam, R.; Chang, J. P., Dielectric property and thermal stability of HfO2 on silicon. *Applied Physics Letters* 2002, 81 (11), 2041-2043.

[13] Tapily, K.; Jakes, J. E.; Stone, D. S.; Shrestha, P.; Gu, D.; Baumgart, H.; Elmustafa, A. A., Nanoindentation Investigation of HfO2 and Al2O3 Films Grown by Atomic Layer Deposition. *Journal of The Electrochemical Society* 2008, 155 (7), H545-H551.

[14] Larsen, E. M., Recent advances in the chemistry of zirconium and hafnium. *Journal of Chemical Education* 1951, 28 (10), 529.

[15] Curtis, C. E.; Doney, L. M.; Johnson, J. R., Some Properties of Hafnium Oxide, Hafnium Silicate, Calcium Hafnate, and Hafnium Carbide. *Journal of the American Ceramic Society* 1954, 37 (10), 458-465.

[16] Adam, J.; Rogers, M. D., The crystal structure of ZrO2 and HfO2. Acta *Crystallographica* 1959, 12 (11), 951-951.

[17] Stacy, D. W.; Johnstone, J. K.; Wilder, D. R., Axial Thermal Expansion of HfO_2. *Journal of the American Ceramic Society* 1972, 55 (9), 482-483.

[18] (a) Smith, D. K.; Cline, C. F., Verification of Existence of Cubic Zirconia at High Temperature. Journal of the American Ceramic Society 1962, 45 (5), 249-250; (b) Wolten, G. M., Diffusionless Phase Transformations in Zirconia and Hafnia. *Journal of the American Ceramic Society* 1963, 46 (9), 418-422.

[19] Rauwel, P.; Rauwel, E.; Persson, C.; Sunding, M. F.; Galeckas, A., One step synthesis of pure cubic and monoclinic HfO2 nanoparticles:

Correlating the structure to the electronic properties of the two polymorphs. *Journal of Applied Physics* 2012, 112 (10), 104107.

[20] Bocquillon, G.; Susse, C.; Vodar, B., Allotropie de l'oxyde d'hafnium sous haute pression. *Rev. Hautes Temp. Refract.* 1968, 5, 247-251.

[21] Maria, J.-P.; Wickaksana, D.; Parette, J.; Kingon, A. I., Crystallization in SiO$_2$–metal oxide alloys. *Journal of Materials Research* 2002, 17, 1571.

[22] Caravaca, M. A.; Casali, R. A., Ab initio localized basis set study of structural parameters and elastic properties of HfO$_2$ polymorphs. *Journal of Physics: Condensed Matter* 2005, 17 (37), 5795.

[23] Arashi, H., Pressure-Induced Phase Transformation of HfO$_2$. *Journal of the American Ceramic Society* 1992, 75 (4), 844-847.

[24] Jayaraman, A.; Wang, S. Y.; Sharma, S. K.; Ming, L. C., Pressure-induced phase transformations in HfO2 to 50 GPa studied by Raman spectroscopy. *Physical Review B* 1993, 48 (13), 9205-9211.

[25] Leger, J. M.; Atouf, A.; Tomaszewski, P. E.; Pereira, A. S., Pressure-induced phase transitions and volume changes in HfO2 up to 50 GPa. *Physical Review B* 1993, 48 (1), 93-98.

[26] Desgreniers, S.; Lagarec, K., High-density ZrO$_2$ and HfO$_2$: Crystalline structures and equations of state. *Physical Review B* 1999, 59 (13), 8467-8472.

[27] Ohtaka, O.; Fukui, H.; Kunisada, T.; Fujisawa, T.; Funakoshi, K.; Utsumi, W.; Irifune, T.; Kuroda, K.; Kikegawa, T., Phase Relations and Volume Changes of Hafnia under High Pressure and High Temperature. *Journal of the American Ceramic Society* 2001, 84 (6), 1369-1373.

[28] Lowther, J. E.; Dewhurst, J. K.; Leger, J. M.; Haines, J., Relative stability of ZrO2 and HfO2 structural phases. *Physical Review B* 1999, 60 (21), 14485-14488.

[29] Lopato, L. M.; Shevchenko, A. V.; Zoz, E. I., *Izv. Akad. Khim.* 1979, 24, 2923.

[30] Delamarre, C.; Perez y Jorba, M., Le system chaux-oxyde de Hafnium. *Rev. Hautes Temp. Refract.* 1965, 2, 313.

[31] Sorolina, S. L.; Scolis, J. Y., Studies of phase relations and stabilities in the HfO2-rich region of the system HfO$_2$-CaO. *Bol. Soc. Esp. Ceram. Vidr.* 1992, 31, 227-231.

[32] López-garcía, A.; Alonso, R. E.; Falabella, M.; Echeverría, G., The Effect of Oxygen Vacancies in Ca1−xSrxHfO$_3$. *Ferroelectrics* 2010, 396 (1), 37-48.

[33] a) Feteira, A.; Sinclair, D. C.; Rajab, K. Z.; Lanagan, M. T., Crystal Structure and Microwave Dielectric Properties of Alkaline-Earth

Hafnates, AHfO$_3$ (A=Ba, Sr, Ca). *Journal of the American Ceramic Society 2008*, 91 (3), 893-901; (b) Ji, Y. M.; Jiang, D. Y.; Wu, Z. H.; Feng, T.; Shi, J. L., Combustion synthesis and photoluminescence of Ce^{3+} activated MHfO$_3$ (M= Ba, Sr, or Ca). *Materials Research Bulletin* 2005, 40 (9), 1521-1526.

[34] a) Rauwel, E.; Galeckas, A.; Rauwel, P.; Fjellvåg, H., Unusual Photoluminescence of CaHfO3 and SrHfO3 Nanoparticles. Advanced Functional Materials 2012, 22 (6), 1174-1179; (b) Rauwel, E.; Galeckas, A.; Rauwel, P.; Wragg, D. S., Response to "Comment on 'Unusual Photoluminescence of CaHfO3 and SrHfO3 Nanoparticles'". *Advanced Functional Materials* 2012, 22 (6), 1114-1115.

[35] Stubican, V. S., Advances in Ceramics, "Science and Technology of Zirconia III". *The American Ceramics Society*, Westerville, OH ed.; 1988; Vol. 24.

[36] Duwez, P.; Brown, F. H.; Odell, F., The Zirconia-Yttria System. *Journal of The Electrochemical Society* 1951, 98 (9), 356-362.

[37] Stacy, D. W.; Wilder, D. R., The Yttria-Hafnia System. *Journal of the American Ceramic Society* 1975, 58 (7-8), 285-288.

[38] Caillet, M.; Deportes, C.; Robert, G.; Vitter, G., Structural study in the system HfO$_2$-Y2O$_3$. *Rev. Int. Hautes Temp. Refract.* 1967, 4, 269-271.

[39] Schieltz, J. D.; Patterson, J. W.; Wilder, D. R., Electrolytic Behavior of Yttria-Stabilized Hafnia. *Journal of The Electrochemical Society* 1971, 118 (8), 1257-1261.

[40] (a) Duran, P.; Pascual, C., Phase equilibria and ordering in the system HfO2-Yb2O3. J. Mater. Sci 1984, 19, 1178-1184; (b) Duran, P.; Pascual, C.; Coutures, J.-P.; Skaggs, S. R., Phase Relations and Ordering in the System Erbia-Hafnia. *Journal of the American Ceramic Society* 1983, 66 (2), 101-106.

[41] Coutures, J. P.; Coutures, J., The System HfO$_2$-TiO$_2$. *Journal of the American Ceramic Society* 1987, 70 (6), 383-387.

[42] Choi, J. H.; Mao, Y.; Chang, J. P., Development of hafnium based high-k materials—A review. *Materials Science and Engineering: R: Reports* 2011, 72 (6), 97-136.

[43] Stemmer, S., Thermodynamic considerations in the stability of binary oxides for alternative gate dielectrics in complementary metal–oxide–semiconductors. *Journal of Vacuum Science & Technology B* 2004, 22 (2), 791-800.

[44] Zhao, X.; Vanderbilt, D., First-principles study of structural, vibrational, and lattice dielectric properties of hafnium oxide. *Physical Review B* 2002, 65 (23), 233106.

[45] a) Cheynet, M. C.; Pokrant, S.; Tichelaar, F. D.; Rouvière, J.-L., Crystal structure and band gap determination of HfO$_2$ thin films. *Journal of Applied Physics* 2007, 101 (5), -; (b) Dubourdieu, C.; Rauwel, E.; Millon, C.; Chaudouët, P.; Ducroquet, F.; Rochat, N.; Rushworth, S.; Cosnier, V., Growth by Liquid-Injection MOCVD and Properties of HfO$_2$ Films for Microelectronic Applications. *Chemical Vapor Deposition* 2006, 12 (2-3), 187-192.

[46] Kim, H.; McIntyre, P. C.; Saraswat, K. C., Effects of crystallization on the electrical properties of ultrathin HfO$_2$ dielectrics grown by atomic layer deposition. *Applied Physics Letters* 2003, 82 (1), 106-108.

[47] Cho, D.-Y.; Jung, H. S.; Yu, I.-H.; Yoon, J. H.; Kim, H. K.; Lee, S. Y.; Jeon, S. H.; Han, S.; Kim, J. H.; Park, T. J.; Park, B.-G.; Hwang, C. S., Stabilization of Tetragonal HfO$_2$ under Low Active Oxygen Source Environment in Atomic Layer Deposition. *Chemistry of Materials* 2012, 24 (18), 3534-3543.

[48] Manory, R. R.; Mori, T.; Shimizu, I.; Miyake, S.; Kimmel, G., Growth and structure control of HfO$_2$ films with cubic and tetragonal structures obtained by ion beam assisted deposition. *Journal of Vacuum Science & Technology A: Vacuum, Surfaces, and Films* 2002, 20 (2), 549-554.

[49] Aarik, J.; Aidla, A.; Kiisler, A. A.; Uustare, T.; Sammelselg, V., Influence of substrate temperature on atomic layer growth and properties of HfO2 thin films. *Thin Solid Films* 1999, 340 (1–2), 110-116.

[50] Aarik, J.; Aidla, A.; Mandar, H.; Uustare, T.; Kukli, K.; Schuisky, M., Phase transformations in hafnium dioxide thin films grown by atomic layer deposition at high temperatures. *Applied Surface Science* 2001, 173 (1-2), 15-21.

[51] Shandalov, M.; McIntyre, P. C., Size-dependent polymorphism in HfO2 nanotubes and nanoscale thin films. *Journal of Applied Physics* 2009, 106 (8), -.

[52] Thomas, R.; Rije, E.; Ehrhart, P.; Milanov, A.; Bhakta, R.; Bauneman, A.; Devi, A.; Fischer, R.; Waser, R., Thin Films of HfO$_2$ for High-k Gate Oxide Applications from Engineered Alkoxide- and Amide-Based MOCVD Precursors. *Journal of The Electrochemical Society* 2007, 154 (3), G77-G84.

[53] a) Hendrix, B. C.; Borovik, A. S.; Wang, Z.; Xu, C.; Roeder, J. F.; Baum, T. H.; Bevan, M. J.; Visokay, M. R.; Chambers, J. J.; Rotondaro,

A. L. P.; Bu, H.; Colombo, L., *Mater. Res. Soc. Symp. Proc.* 2002, 716, B.6.7.1; (b) Van Elshocht, S.; Caymax, M.; DeGendt, S.; Conrad, T.; Petry, J.; Claes, M.; Witters, T.; Zhao, C.; Brijs, B.; Richard, O.; Bender, H.; Vandervorst, W.; Carter, R.; Kluth, J.; Date, L.; Pique, D.; Heyns, M. M., *Mater. Res. Soc. Symp. Proc.* 2003, 745, N.5.15.5.

[54] Aarik, J.; Aidla, A.; Mändar, H.; Sammelselg, V.; Uustare, T., Texture development in nanocrystalline hafnium dioxide thin films grown by atomic layer deposition. *Journal of Crystal Growth* 2000, 220 (1–2), 105-113.

[55] Hiroya, I.; Satoru, G.; Kazutaka, H.; Mitsuo, S.; Akira, S.; Shigeaki, Z.; Yukio, Y., Structural and Electrical Characteristics of HfO_2 Films Fabricated by Pulsed Laser Deposition. *Japanese Journal of Applied Physics* 2002, 41 (4S), 2476.

[56] Wang, H.; Wang, Y.; Zhang, J.; Ye, C.; Wang, H. B.; Feng, J.; Wang, B. Y.; Li, Q.; Jiang, Y., Interface control and leakage current conduction mechanism in HfO_2 film prepared by pulsed laser deposition. *Applied Physics Letters* 2008, 93, 202904.

[57] He, G.; Liu, M.; Zhu, L. Q.; Chang, M.; Fang, Q.; Zhang, L. D., Effect of postdeposition annealing on the thermal stability and structural characteristics of sputtered HfO2 films on Si (100). *Surface Science* 2005, 576 (1–3), 67-75.

[58] Pereira, L.; Barquinha, P.; Fortunato, E.; Martins, R., Influence of the oxygen/argon ratio on the properties of sputtered hafnium oxide. *Materials Science and Engineering: B* 2005, 118 (1–3), 210-213.

[59] Migita, S.; Watanabe, Y.; Ota, H.; Ito, H.; Kamimuta, Y.; Nabatame, T.; Toriumi, A. In Design and Demonstration of Very High-k (k~50) HfO_2 for Ultra-Scaled Si CMOS, Symposium on VLSI Technology, 2008; pp 152-153.

[60] Li, F. M.; Bayer, B. C.; Hofmann, S.; Dutson, J. D.; Wakeham, S. J.; Thwaites, M. J.; Milne, W. I.; Flewitt, A. J., High-k (k=30) amorphous hafnium oxide films from high rate room temperature deposition. *Applied Physics Letters* 2011, 98, 252903.

[61] a) Ervin, G., Structural interpretation of the diaspore–corundum and boehmite–γ-Al2O3 transitions. *Acta Crystallographica* 1952, 5 (1), 103-108; (b) Wilson, S. J., The dehydration of boehmite, γ-AlOOH, to γ-Al_2O_3. *Journal of Solid State Chemistry* 1979, 30 (2), 247-255.

[62] a) Rauwel, E.; Rauwel, P.; Ducroquet, F.; Matko, I.; Lourenço, A. C., Metallic oxygen barrier diffusion applied to high-κ deposition. *Journal of Vacuum Science & Technology B 2011*, 29, 01A701; (b) Rauwel,

E.; Rauwel, P.; Ducroquet, F.; Sunding, M. F.; Matko, I.; Lourenço, A. C., Magnesium metallic interlayer as an oxygen-diffusion-barrier between high-κ dielectric thin films and silicon substrate. *Thin Solid Films* 2012, 520 (17), 5602-5609.

[63] a) Rauwel, P.; Løvvik, O. M.; Rauwel, E.; Taftø, J., Nanovoids in thermoelectric β-Zn4Sb3: A possibility for nanoengineering via Zn diffusion. *Acta Materialia* 2011, 59 (13), 5266-5275; (b) Smigelskas, A. D.; Kirkendall, E. O., Zinc Diffusion in Alpha Brass. *Trans. AIME* 1947, 171 ( ), 130-142.

[64] a) Leskelä, M.; Kemell, M.; Kukli, K.; Pore, V.; Santala, E.; Ritala, M.; Lu, J., Exploitation of atomic layer deposition for nanostructured materials. *Materials Science and Engineering: C* 2007, 27 (5–8), 1504-1508; (b) Rauwel, E.; Nilsen, O.; Rauwel, P.; Walmsley, J. C.; Frogner, H. B.; Rytter, E.; Fjellvåg, H., *Oxide Coating of Alumina Nanoporous Structure Using ALD to Produce Highly Porous Spinel. Chemical Vapor Deposition* 2012, 18 (10-12), 315-325.

[65] Takahashi, K.; Nakayama, M.; Hino, S.; Tokumitsu, E.; Funakubo, H., Growth of Hafnium Oxide Films by Metalorganic Chemical Vapor Deposition Using Oxygen-Free Hf[N(C2H5)2]4 Precursor and Their Properties. *Integrated Ferroelectrics* 2003, 57 (1), 1185-1192.

[66] a) Loo, Y. F.; O'Kane, R.; Jones, A. C.; Aspinall, H. C.; Potter, R. J.; Chalker, P. R.; Bickley, J. F.; Taylor, S.; Smith, L. M., Deposition of HfO2 and ZrO2 films by liquid injection MOCVD using new monomeric alkoxide precursors. *Journal of Materials Chemistry* 2005, 15 (19), 1896-1902; (b) Loo, Y. F.; O'Kane, R.; Jones, A. C.; Aspinall, H. C.; Potter, R. J.; Chalker, P. R.; Bickley, J. F.; Taylor, S.; Smith, L. M., Deposition of HfO$_2$ Films by Liquid Injection MOCVD Using a New Monomeric Alkoxide Precursor, [Hf(dmop)4]. *Chemical Vapor Deposition* 2005, 11 (6-7), 299-305.

[67] a) Van Elshocht, S.; Baklanov, M.; Brijs, B.; Carter, R.; Caymax, M.; Carbonell, L.; Claes, M.; Conard, T.; Cosnier, V.; Daté, L.; De Gendt, S.; Kluth, J.; Pique, D.; Richard, O.; Vanhaeren, D.; Vereecke, G.; Witters, T.; Zhao, C.; Heyns, M., Bulk Properties of MOCVD-Deposited HfO$_2$ Layers for High k Dielectric Applications. *Journal of The Electrochemical Society 2004*, 151 (10), F228-F234; (b) Gould, B. J.; Povey, I. M.; Pemble, M. E.; Flavell, W. R., Chemical vapour deposition of ZrO$_2$ thin films monitored by IR spectroscopy. *Journal of Materials Chemistry* 1994, 4 (12), 1815-1819.

[68] Schaeffer, J.; Edwards, N. V.; Liu, R.; Roan, D.; Hradsky, B.; Gregory, R.; Kulik, J.; Duda, E.; Contreras, L.; Christiansen, J.; Zollner, S.; Tobin, P.; Nguyen, B.-Y.; Nieh, R.; Ramon, M.; Rao, R.; Hegde, R.; Rai, R.; Baker, J.; Voight, S., HfO2 Gate Dielectrics Deposited via Tetrakis Diethylamido Hafnium. *Journal of The Electrochemical Society* 2003, 150 (4), F67-F74.

[69] Jones, A. C.; Aspinall, H. C.; Chalker, P. R.; Potter, R. J.; Manning, T. D.; Loo, Y. F.; O'Kane, R.; Gaskell, J. M.; Smith, L. M., MOCVD and ALD of High-k Dielectric Oxides Using Alkoxide Precursors. *Chemical Vapor Deposition* 2006, 12 (2-3), 83-98.

[70] Teren, A. R.; Ehrhart, P.; Waser, R.; He, J. Q.; Jia, C. L.; Schumacher, M.; Lindner, J.; Baumann, P. K.; Leedham, T. J.; Rushworth, S. R.; Jones, A. C., Comparison of Hafnium Precursors for the MOCVD of HfO$_2$ for Gate Dielectric Applications. *Integrated Ferroelectrics* 2003, 57 (1), 1163-1173.

[71] Dubourdieu, C.; Roussel, H.; Jimenez, C.; Audier, M.; Sénateur, J. P.; Lhostis, S.; Auvray, L.; Ducroquet, F.; O'Sullivan, B. J.; Hurley, P. K.; Rushworth, S.; Hubert-Pfalzgraf, L., Pulsed liquid-injection MOCVD of high-K oxides for advanced semiconductor technologies. *Materials Science and Engineering: B* 2005, 118 (1–3), 105-111.

[72] Rauwel, E. e. a., unpublished work.

[73] Kang, H.; Kim, S.; Choi, J.; Kim, J.; Jeon, H.; Bae, C., Effects of Remote Plasma Pre-oxidation of Si Substrates on the Characteristics of ALD-Deposited HfO2 Gate Dielectrics. *Electrochemical and Solid-State Letters* 2006, 9 (6), G211-G214.

[74] a) Sreenivasan, R.; McIntyre, P. C.; Kim, H.; Saraswat, K. C., Effect of impurities on the fixed charge of nanoscale HfO$_2$ films grown by atomic layer deposition. *Applied Physics Letters* 2006, 89 (11), -; (b) Cho, M.; Park, H. B.; Park, J.; Lee, S. W.; Hwang, C. S.; Jang, G. H.; Jeong, J., High-k properties of atomic-layer-deposited HfO$_2$ films using a nitrogen-containing Hf[N(CH3)2]4 precursor and H$_2$O oxidant. *Applied Physics Letters* 2003, 83 (26), 5503-5505; (c) Kukli, K.; Ritala, M.; Leskelä, M.; Sajavaara, T.; Keinonen, J.; Jones, A. C.; Roberts, J. L., Atomic Layer Deposition of Hafnium Dioxide Films Using Hafnium Bis(2-butanolate)bis(1-methoxy-2-methyl-2-propanolate) and Water. *Chemical Vapor Deposition* 2003, 9 (6), 315-320; (d) Rauwel, E.; Willinger, M.-G.; Ducroquet, F.; Rauwel, P.; Matko, I.; Kiselev, D.; Pinna, N., Carboxylic Acids as Oxygen Sources for the Atomic Layer Deposition of High-κ Metal Oxides. *The Journal of Physical Chemistry*

C 2008, 112 (33), 12754-12759; (e) Conley, J. F.; Ono, Y.; Tweet, D. J.; Zhuang, W.; Solanki, R., Atomic layer deposition of thin hafnium oxide films using a carbon free precursor. *Journal of Applied Physics* 2003, 93 (1), 712-718.

[75]   a) Aarik, J.; Aidla, A.; Mändar, H.; Uustare, T.; Kukli, K.; Schuisky, M., Phase transformations in hafnium dioxide thin films grown by atomic layer deposition at high temperatures. *Applied Surface Science* 2001, 173 (1–2), 15-21; (b) Conley, J. F.; Ono, Y.; Zhuang, W.; Tweet, D. J.; Gao, W.; Mohammed, S. K.; Solanki, R., Atomic Layer Deposition of Hafnium Oxide Using Anhydrous Hafnium Nitrate. *Electrochemical and Solid-State Letters* 2002, 5 (5), C57-C59; (c) Kim, S.; Kim, J.; Choi, J.; Kang, H.; Jeon, H.; Cho, W.; An, K.-S.; Chung, T.-M.; Kim, Y.; Bae, C., Remote Plasma Atomic Layer Deposition of HfO$_2$ Thin Films Using the Alkoxide Precursor Hf(mp)4. *Electrochemical and Solid-State Letters* 2006, 9 (6), G200-G203; (d) Kukli, K.; Ritala, M.; Lu, J.; Haårsta, A.; Leskelä, M., Properties of HfO2 Thin Films Grown by ALD from Hafnium tetrakis(ethylmethylamide) and Water. *Journal of The Electrochemical Society* 2004, 151 (8), F189-F193.

[76]   Ho, M.-Y.; Gong, H.; Wilk, G. D.; Busch, B. W.; Green, M. L.; Voyles, P. M.; Muller, D. A.; Bude, M.; Lin, W. H.; See, A.; Loomans, M. E.; Lahiri, S. K.; Räisänen, P. I., Morphology and crystallization kinetics in HfO$_2$ thin films grown by atomic layer deposition. *Journal of Applied Physics* 2003, 93 (3), 1477-1481.

[77]   Kim, J. H.; Park, T. J.; Cho, M.; Jang, J. H.; Seo, M.; Na, K. D.; Hwang, C. S.; Won, J. Y., Reduced Electrical Defects and Improved Reliability of Atomic-Layer-Deposited HfO$_2$ Dielectric Films by In Situ NH3 Injection. *Journal of The Electrochemical Society* 2009, 156 (5), G48-G52.

[78]   (a) Nishide, T.; Honda, S.; Matsuura, M.; Ide, M., Surface, structural and optical properties of sol-gel derived HfO$_2$ films. *Thin Solid Films* 2000, 371 (1–2), 61-65; (b) Blanchin, M. G.; Canut, B.; Lambert, Y.; Teodorescu, V. S.; Barău, A.; Zaharescu, M., Structure and dielectric properties of HfO$_2$ films prepared by a sol–gel route. *J Sol-Gel Sci Technol* 2008, 47 (2), 165-172.

[79]   Navrotsky, A., Thermochemical insights into refractory ceramic materials based on oxides with large tetravalent cations. *Journal of Materials Chemistry* 2005, 15 (19), 1883-1890.

[80]   (a) Rauwel, E.; Dubourdieu, C.; Holländer, B.; Rochat, N.; Ducroquet, F.; Rossell, M. D.; Van Tendeloo, G.; Pelissier, B., Stabilization of the

cubic phase of HfO_2 by Y addition in films grown by metal organic chemical vapor deposition. *Applied Physics Letters* 2006, 89, 012902; (b) Dai, J. Y.; Lee, P. F.; Wong, K. H.; Chan, H. L. W.; Choy, C. L., Epitaxial growth of yttrium-stabilized HfO_2 high-k gate dielectric thin films on Si. *Journal of Applied Physics* 2003, 94 (2), 912-915; (c) Pasko, S. V.; Hubert-Pfalzgraf, L. G.; Abrutis, A.; Richard, P.; Bartasyte, A.; Kazlauskiene, V., New sterically hindered Hf, Zr and Y [small beta]-diketonates as MOCVD precursors for oxide films. *Journal of Materials Chemistry* 2004, 14 (8), 1245-1251; (d) Kita, K.; Kyuno, K.; Toriumi, A., Permittivity increase of yttrium-doped HfO_2 through structural phase transformation. *Applied Physics Letters* 2005, 86 (10), -; (e) Tao, Q.; Jursich, G.; Majumder, P.; Singh, M.; Walkosz, W.; Gu, P.; Klie, R.; Takoudis, C., Composition–Structure–Dielectric Property of Yttrium-Doped Hafnium Oxide Films Deposited by Atomic Layer Deposition. *Electrochemical and Solid-State Letters* 2009, 12 (9), G50-G53; (f) Niinistö, J.; Kukli, K.; Sajavaara, T.; Ritala, M.; Leskelä, M.; Oberbeck, L.; Sundqvist, J.; Schröder, U., Atomic Layer Deposition of High-Permittivity Yttrium-Doped HfO2 Films. *Electrochemical and Solid-State Letters* 2009, 12 (1), G1-G4; (g) Yang, Z. K.; Lee, W. C.; Lee, Y. J.; Chang, P.; Huang, M. L.; Hong, M.; Hsu, C.-H.; Kwo, J., Cubic HfO_2 doped with Y2O3 epitaxial films on GaAs (001) of enhanced dielectric constant. *Applied Physics Letters* 2007, 90 (15), 152908; (h) Komatsu, M.; Yasuhara, R.; Takahashi, H.; Toyoda, S.; Kumigashira, H.; Oshima, M.; Kukuruznyak, D.; Chikyow, T., Crystal structures and band offsets of ultrathin $HfO_2–Y_2O_3$ composite films studied by photoemission and x-ray absorption spectroscopies. *Applied Physics Letters* 2006, 89, 172107; (i) Wang, X. F.; Li, Q.; Moreno, M. S., Effect of Al and Y incorporation on the structure of HfO_2. *Journal of Applied Physics* 2008, 104, 093529.

[81] Dubourdieu, C.; Rauwel, E.; Roussel, H.; Ducroquet, F.; Holländer, B.; Rossell, M.; Van Tendeloo, G.; Lhostis, S.; Rushworth, S., Addition of yttrium into HfO_2 films: Microstructure and electrical properties. *Journal of Vacuum Science & Technology A: Vacuum, Surfaces, and Films* 2009, 27 (3), 503.

[82] a) Yakovkina, L. V.; Smirnova, T. P.; Borisov, V. O.; Jeong-Hwan, S.; Morozova, N. B.; Kichai, V. N.; Smirnov, A. V., Structure and properties of films based on HfO_2-Sc_2O_3 double oxide. *J Struct Chem* 2011, 52 (4), 743-747; (b) Yakovkina, L. V.; Smirnova, T. P.; Borisov, V. O.; Kichai, V. N.; Kaichev, V. V., Synthesis and properties of

dielectric (HfO$_2$)1 − x (Sc$_2$O$_3$) x films. *Inorg Mater* 2013, 49 (2), 172-178.

[83] Brize, V.; Terrenoir, F.; Matko, I.; Rochat, N.; Holländer, B.; Feist, B.; Blasco, N.; Dubourdieu, C., Sc Addition In HfO$_2$ Thin Films Prepared By Liquid-injection MOCVD. *ECS Transactions* 2008, 13 (1), 157-162.

[84] (a) Neumayer, D. A.; Cartier, E., Materials characterization of ZrO$_2$–SiO$_2$ and HfO$_2$–SiO$_2$ binary oxides deposited by chemical solution deposition. *Journal of Applied Physics* 2001, 90 (4), 1801-1808; (b) Kim, H.; McIntyre, P. C., Spinodal decomposition in amorphous metal–silicate thin films: Phase diagram analysis and interface effects on kinetics. *Journal of Applied Physics* 2002, 92 (9), 5094-5102; (c) Stemmer, S.; Li, Y.; Foran, B.; Lysaght, P. S.; Streiffer, S. K.; Fuoss, P.; Seifert, S., Grazing-incidence small angle x-ray scattering studies of phase separation in hafnium silicate films. *Applied Physics Letters* 2003, 83 (15), 3141-3143.

[85] Ushakov, S. V.; Navrotsky, A.; Yang, Y.; Stemmer, S.; Kukli, K.; Ritala, M.; Leskelä, M. A.; Fejes, P.; Demkov, A.; Wang, C.; Nguyen, B. Y.; Triyoso, D.; Tobin, P., *Crystallization in hafnia- and zirconia-based systems. physica status solidi* (b) 2004, 241 (10), 2268-2278.

[86] Toriumi, A.; Kita, K.; Tomida, K.; Yamamoto, Y., Doped HfO2 for Higher-k Dielectrics. ECS Transactions 2006, 1 (5), 185-197.

[87] Yamamoto, Y.; Kita, K.; Kyuno, K.; Toriumi, A., Structural and electrical properties of HfLaOx films for an amorphous high-k gate insulator. *Applied Physics Letters* 2006, 89 (3), 032903.

[88] a) Adelmann, C.; Sriramkumar, V.; Van Elshocht, S.; Lehnen, P.; Conard, T.; De Gendt, S., Dielectric properties of dysprosium- and scandium-doped hafnium dioxide thin films. *Applied Physics Letters* 2007, 91, 162902; (b) Losovyj, Y. B.; Ketsman, I.; Sokolov, A.; Belashchenko, K. D.; Dowben, P. A.; Tang, J.; Wang, Z., The electronic structure change with Gd doping of HfO$_2$ on silicon. Applied Physics Letters 2007, 91, 132908; (c) Govindarajan, S.; Böscke, T. S.; Sivasubramani, P.; Kirsch, P. D.; Lee, B. H.; Tseng, H.-H.; Jammy, R.; Schröder, U.; Ramanathan, S.; Gnade, B. E., Higher permittivity rare earth doped HfO$_2$ for sub-45-nm metal-insulator-semiconductor devices. *Applied Physics Letters* 2007, 91, 062906.

[89] (a) Rauwel, E., unpublished work; (b) Ducroquet, F.; Rauwel, E.; Brizé, V.; Dubourdieu, C., (Invited) Dielectric Properties and Flat-Band Voltages of Doped- HfO$_2$ Thin Films. *ECS Transactions* 2010, 28 (2), 191-202.

[90] (a) Yu, H. Y.; Li, M. F.; Cho, B. J.; Yeo, C. C.; Joo, M. S.; Kwong, D.-L.; Pan, J. S.; Ang, C. H.; Zheng, J. Z.; Ramanathan, S., Energy gap and band alignment for $(HfO_2)x(Al_2O_3)1-x$ on (100) Si. *Applied Physics Letters* 2002, 81 (2), 376-378; (b) Wang, X. F.; Li, Q.; Egerton, R. F.; Lee, P. F.; Dai, J. Y.; Hou, Z. F.; Gong, X. G., Effect of Al addition on the microstructure and electronic structure of HfO_2 film. *Journal of Applied Physics* 2007, 101 (1), 013514.

[91] Lu, J.; Kuo, Y.; Tewg, J.-Y., Hafnium-Doped Tantalum Oxide High-k Gate Dielectrics. *Journal of The Electrochemical Society* 2006, 153 (5), G410-G416.

[92] Yu, X.; Zhu, C.; Li, M. F.; Chin, A.; Du, A. Y.; Wang, W. D.; Kwong, D.-L., Electrical characteristics and suppressed boron penetration behavior of thermally stable HfTaO gate dielectrics with polycrystalline-silicon gate. *Applied Physics Letters* 2004, 85 (14), 2893-2895.

[93] (a) Garvie, R. C., Stabilization of the tetragonal structure in zirconia microcrystals. *The Journal of Physical Chemistry* 1978, 82 (2), 218-224; (b) Filipovich, V.; Kalinini, A., Struct. *Glass.* 1965, 34, 34.

[94] a) Mazdiyasni, K. S.; Lynch, C. T.; Smith, J. S., Preparation of Ultra-High-Purity Submicron Refractory Oxides. *Journal of the American Ceramic Society 1965*, 48 (7), 372-375; (b) Mazdiyasni, K. S.; Brown, L. M., Preparation and Characterization of Submicron Hafnium Oxide. *Journal of the American Ceramic Society* 1970, 53 (1), 43-45.

[95] Tirosh, E.; Markovich, G., Control of Defects and Magnetic Properties in Colloidal HfO_2 Nanorods. *Advanced Materials* 2007, 19 (18), 2608-2612.

[96] Toraya, H.; Yoshimura, M.; Somiya, S., Hydrothermal Reaction-Sintering of Monoclinic HfO_2. *Journal of the American Ceramic Society* 1982, 65 (9), c159-c160.

[97] Eliziario, S.; Cavalcante, L.; Sczancoski, J.; Pizani, P.; Varela, J.; Espinosa, J.; Longo, E., Morphology and Photoluminescence of HfO_2 Obtained by Microwave-Hydrothermal. *Nanoscale Research Letters* 2009, 4 (11), 1371 - 1379.

[98] Štefanić, G.; Molčanov, K.; Musić, S., A comparative study of the hydrothermal crystallization of HfO_2 using DSC/TG and XRD analysis. *Materials Chemistry and Physics* 2005, 90 (2–3), 344-352.

[99] Pinna, N.; Garnweitner, G.; Antonietti, M.; Niederberger, M., Non-Aqueous Synthesis of High-Purity Metal Oxide Nanopowders Using an Ether Elimination Process. *Advanced Materials* 2004, 16 (23-24), 2196-2200.

[100] Rauwel, E.; Galeckas, A.; Rauwel, P., Photoluminescent cubic and monoclinic HfO$_2$ nanoparticles: effects of temperature and ambient. *Materials Research Express* 2014, 1, 015035.

[101] a) Brown, L. M.; Mazdiyasni, K. S., Characterization of Alkoxy-Derived Yttria-tabilized Hafnia. *Journal of the American Ceramic Society* 1970, 53 (11), 590-594; (b) Mazdiyasni, K. S.; Lynch, C. T.; Ii, J. S. S., Cubic Phase Stabilization of Translucent Yttria-Zirconia at Very Low Temperatures. *Journal of the American Ceramic Society* 1967, 50 (10), 532-537.

[102] Pucci, A.; Clavel, G.; Willinger, M.-G.; Zitoun, D.; Pinna, N., Transition Metal-Doped ZrO$_2$ and HfO$_2$ Nanocrystals. *The Journal of Physical Chemistry C 2009*, 113 (28), 12048-12058.

[103] (a) Taniguchi, T.; Sakamoto, N.; Watanabe, T.; Matsushita, N.; Yoshimura, M., Rational Hydrothermal Route to Monodisperse Hf1-xEuxO2-x/2 Solid Solution Nanocrystals. The Journal of Physical Chemistry C 2008, 112 (13), 4884-4891; (b) Wiatrowska, A.; Zych, E.; Kępiński, L., Monoclinic HfO$_2$:Eu X-ray phosphor. *Radiation Measurements* 2010, 45 (3–6), 493-496.

[104] Lauria, A.; Villa, I.; Fasoli, M.; Niederberger, M.; Vedda, A., Multifunctional Role of Rare Earth Doping in Optical Materials: Nonaqueous Sol–Gel Synthesis of Stabilized Cubic HfO$_2$ Luminescent Nanoparticles. *ACS Nano* 2013, 7 (8), 7041-7052.

In: Hafnium
Editor: HongYu Yu

ISBN: 978-1-63463-164-8
© 2015 Nova Science Publishers, Inc.

Chapter 5

HAFNIUM-BASED THIN OXIDES: VERSATILE INSULATORS FOR MICROELECTRONICS

***Albin Bayerl and Mario Lanza**[*]*

Institute of Functional Nano and Soft Materials (FUNSOM), Jiangsu
Key Laboratory for Carbon-Based Functional Materials and Devices and
Collaborative Innovation Center of Suzhou Nano Science and Technology,
Soochow University, Suzhou, China

ABSTRACT

Hafnium-based oxides are becoming more and more popular in the field of memory technology. Apart from their proved ability of reducing the leakage current in transistors and nanomemories, HfO2 and related oxides present an almost unique property: Resistive Switching, which makes them a promising candidate for building up contemporary RAM (Random Access Memory) cells. In such cells, information thereof is not recorded by storing charge, but the memory point relies on resistivity changes. A low resistive state (LRS) and high resistive state (HRS) of a conductive filament (CF) will be interpreted as logic '0' and '1', respectively. Resistive-RAM is especially interesting as it presents excellent compatibility with yet established fabrication technology and production lines, and good electrical performance. For this reason great

---

[*] Corresponding Author address: Email: mlanza@suda.edu.cn

 Albin Bayerl and Mario Lanza

investigative effort has been undertaken during the last years. However, in the most promising cells, resistive switching occurs at areas in the ~100nm² range, a fact that makes necessary the use of tools with high lateral resolution. Here, we prove that CAFM can be very useful to get a deeper insight into the process of forming, set and reset. We also demonstrate that RS is a local phenomenon restricted to grain boundaries in polycrystalline layers owing to their electrically weaker behavior as a consequence of defect accumulation.

Keywords: Hafnium oxide, high-k; resistive switching; RRAM; NVM; nanoscale; CAFM; oxygen vacancies; conductive filament; dielectric breakdown

INTRODUCTION

With time, investigation activities to which HfO2 was subjected revealed another astonishing property of this material: the ability of tuning its electrical resistance depending on the voltages applied. This phenomenon has supposed a revolution in the field of Non Volatile Memories (NVM), leading to high performance Resistive Random Access Memory (RRAM). To better understand this breakthrough, it is necessary to look back to 1964 when Arnold Farber and Eugene Schlig, both working for IBM, created the first electronic memory cell. Since the very beginning, memories have stored each bit of data in a single capacitor. The two possible states of a capacitor – "charged" and "uncharged" can be taken to represent the two states of a bit, a "0" and a "1". This is the working principle of the omnipresent flash NAND memory that we use in our USB and SD cards. But since the capacitors will discharge slowly due to current leak, the stored information fades. Now, for the first time and thanks to the RRAM, we can store information in a different and more efficient way. RRAM memories are not volatile because the information is not represented by a charge state, but by very localized and reversible physical changes of its bulk structure. This leads to different states of resistance LRS (Low Resistive State) and HRS (High Resistive State). No matter if the device is powered or not, information, until actively removed, will not vanish. A commercial launch of the first RRAM devices is expected in 2015 [10].

In this work we review the performance of Hafnium-based oxides as dielectric for microelectronic devices. We analyze the electrical performance of thin HfO2 films using the conductive atomic force microscope and related setups. More specifically, we assess the physical origin of resistive switching in these layers and link it to the morphology of the samples.

RESISTIVE SWITCHING IN HIGH-K THIN FILMS

The core of resistance-change memory devices consists of two electrodes sandwiching a thin stack of a material altering its electrical resistance. Transition metal oxides (TMO) play an important role in the group of materials with alterable conductivity, ranging from perovskites like $SrTiO_3$ to binary oxides (composed of two different atoms) such as NiO, including high-k materials [11-13] and HfO_2 [14]. Depending on the physical origin of the switching in these materials, two subgroups can be formed. In the first subgroup the mechanism for the switching has a distributed nature, i. e. the change is formed by the creation of a Schottky barrier [15-16, 17], where charges are driven by an electrical field to the electrodes of the cell. Another approach is based on the formation and destruction of conductive filaments (CF) between the two electrodes [18], an event which depends on the applied voltage.

As the electrical stress proceeds, a critical density of defects establishes an electrical connection between the electrodes by means of a conduction path within the dielectric, leading to a sudden increase of the tunneling current (percolation model [19-21]): this is the onset of the BD and in the field of RRAM is usually referred as CF electroforming (forming process). The percolation model also predicts that BD, although being a randomly distributed phenomenon throughout the respective device, is highly localized to areas in the range of 100nm2. Initially it was believed that the BD was irreversible. This theory later on was to be falsified by subsequent investigation, observing thatunder specific polarization conditions, the dielectric behavior could be partially recovered (reset process). If forming and reset can be achievedby applying a voltage of the same polarity, the RS is called unipolar. On the contrary it is called bipolar [22]. Interestingly, some materials can exhibit both types of RS [23]. In those cases, the key factor determining the presence of one RS mechanism, or both simultaneously is the current level used to form the CF. At low CLs a thin CF is formed by defects that migrate from anode to cathode [24]. These CFs can be easily reoxidized at their narrower end by applying a reverse voltage, leading to bipolar RS [22]. If the current used during the forming process is too high, the CF cannot be reoxidized, which means an irreversible BD. Additionally, CFs may contain impurities from the metallic electrode layers; in this case, the only way to recover the stack dielectric properties is fusing the CF by Joule heat, i. e. by driving higher currents [22,25-26]. This, however, implies a higher power consumption, for which researchers are tending to pay more attention to bipolar RS. The nature of both switching mechanisms can be schematically observed in Figure 1. Due to such lower currents driven at LRS and therefore a lower power consumption, in this work we will focus on bipolar RS.

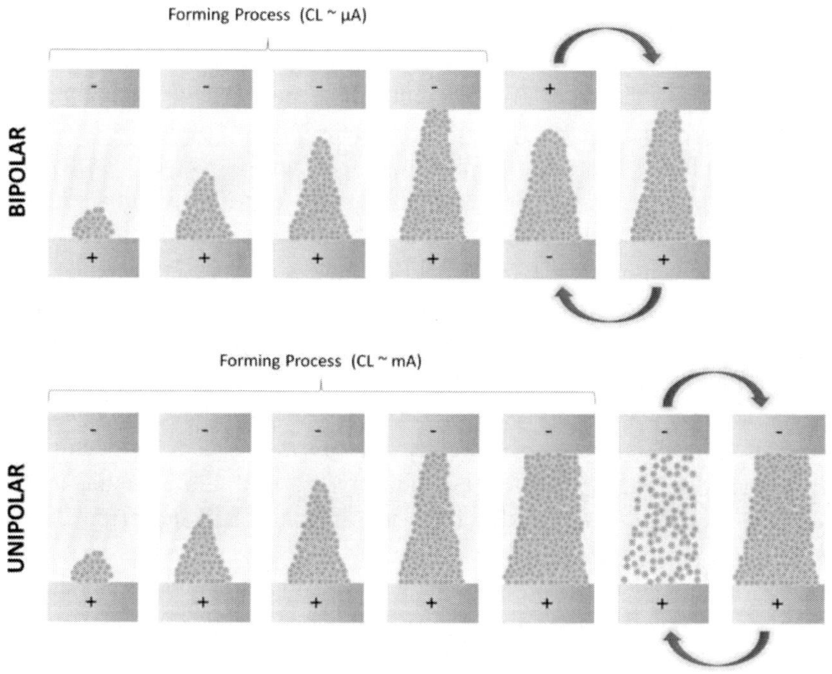

Figure 1. Comparative schematic between bipolar and unipolar resistive switching mechanisms (reoxidation for bipolar and thermal heat for unipolar). The blue and yellow stacks represent the metallic electrodes and insulator (respectively), and the pink circles represent the particles that form the conducting filament. Reproduced with permission from [27]. Published by Materials [2014].

The observation of RS has been traditionally performed by applying electrical stresses in a squared MIM structure of micrometric dimensions using a semiconductor parameter analyzer (SPA) and a probe station. Following this methodology, valuable information about the different properties of the device performance has been gained. Lee et al. [28] fabricated nanofilament nickel oxide channels across two platinum electrodes by applying electrical stress.

They were able to observe a reduction in the switching time as the intersection area decreased, which was attributed to the formation of thinner or weaker filaments. As less switching power is needed, nanofilament-based memory is a promising candidate for low-power applications. Normally, electroforming processes are carried out including current limitation in order to prevent permanent electrical breakdown of the device and to control the current level of the LRS and HRS, as well. However, CeOx-based RRAM device reported by Sun et al. [29] show uniform reset and set voltages excluding a external current limitation. Furthermore, they report bipolar RS behavior. Such devices show different performance due to multiple forming of conductive filaments. This will linger on until a device is fabricated with only one stable conducting filament, which may lead to a great reproducibility and lower power consumption. Resistive memory switching behavior depending on voltage sweep direction was studied by Sung et al. [30]. He created a device by inserting a reactive Ti layer between the TiO_2 insulator and the top electrode. This procedure led to a a device with a more stable counterclockwise directional bipolar switching. Despite these achievements, due to the local nature of the CF, the physical origin of the switching mechanism has never been totally understood.

NANOSCALE ELECTRICAL CHARACTERIZATION

As mentioned above, the area of the reversible conductive filament responsible of the RS ranges roughly from 1 to 100 nm^2. In order to provide in-situ information about the CF forming process, electronic characterization tools providing a higher resolution are necessary. The most widespread methodologies in the field of nanoscale electrical characterization are Scanning Tunneling Microscopy (STM) and Conductive AFM (CAFM). In the case of STM, despite a conductive sample is needed, research groups have managed to measure changes in local conduction on insulating films. The group headed by K. L. Pey [31] collected IV curves and current maps on the surface of polycrystalline CeO_2, reporting that the current through the grain boundaries was larger. Kwon et al. [32] even combined STM with TEM to investigate the structure of conductive nanofilaments in TiO_2 based memory cells that show cyclic resistance changes. Nonetheless, current measurements by STM-based technologies correspond to the tip/sample tunneling current by maintaining the tip separated just a few nanometers from its surface. Effectively, it is not in physical contact with the sample implying an additive resistive component to the tip-sample arrangement; this drawback is avoided when using the CAFM working in contact mode since it can simultaneously analyze the topography and electrical properties of thin oxides. CAFM is already being used now for more than two decades, and it has proved to be a valuable tool to investigate many different kinds of materials – not only conductors. For example, CAFM experiments allowed to detect that high-k materials present an excess of defects compared to SiO_2, which consequently promotes an increase in leakage current [33]. In Hafnium dioxide layers, the CAFM has helped to reveal how the electrical properties of the stack can change depending on thickness fluctuations, amount of defects and stress applied [34, 35]. In the field of RS, the CAFM has been also successfully used to assess many interesting phenomena such as the density of CFs in a MIM structure in HRS and LRS, as well as their size and currents driven. Such information can be obtained by performing two different experiments with a commercial CAFM.

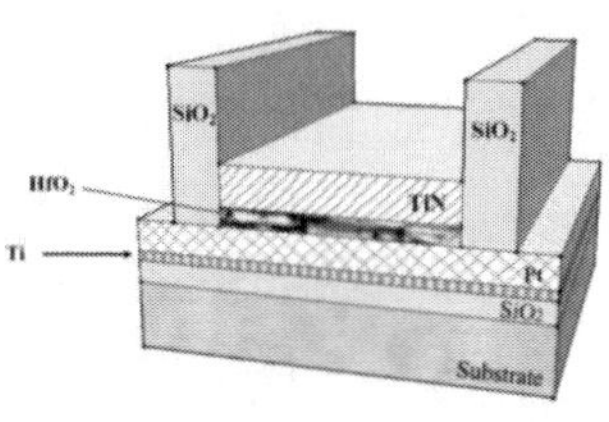
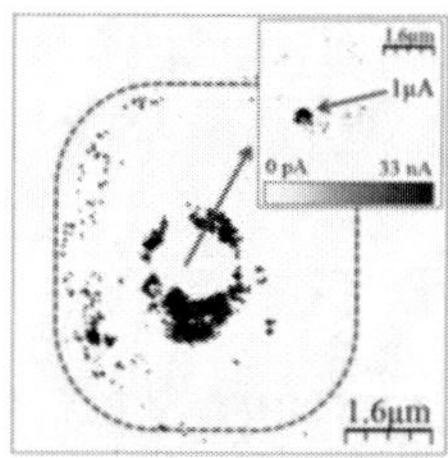
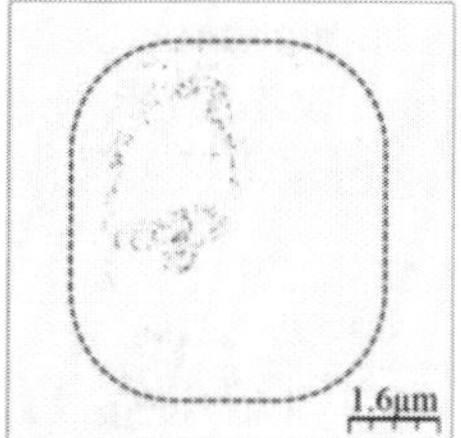

Figure 2. (a) Schematic of a TiN/HfO$_2$/Pt/Ti/SiO$_2$/Si memory cell. HRS and LRS are induced at device level. The top Al contacts are removed and the underlying insulator is investigated with CAFM biasing the tip. (b) and (c) show current maps of a capacitor in HRS (after an erasing operation) and the other in LRS (after a write operation), respectively. Reproduced with permission from Elsevier B.V. Copyright [2012].

The first one uses a stack where a dielectric material is sandwiched between two electrodes. Typically, the resistance change is recorded in a MIM structure and subsequently, the contact electrode is removed chemically and the bare insulating material is scanned by means of CAFM. Iglesias et al. [36] used polycrystalline HfO$_2$ layer contacted by TiN at the top and with a Pt electrode at the bottom to investigate the resistive-switching phenomena (Figure 2). The HRS and LRS is directly shown by CAFM current images obtained by scanning on the respective cells' insulating layer (after top electrode removal). Its superior resolution has shown that the amount of leaky sites in the LRS structure was larger implying a smaller respectiveresistance. It has to be noted that a voltage was applied in order to simulate a read-out voltage. Under normal operation conditions this voltage would be applied uniformly throughout the entire device. This experiment afterwards was extended onto different materials to determine their RS behavior, such as silicon nitride based dielectrics.

Kim et al. [37] used a memory cell made up of $Ag/Si_3N_4/Al$ stack. CAFM results reveal that the respective leaky site distributions related to the different resistive states are comparable with the current maps of [36]. Additionally, they found out that the injected charges compared to the changes in resistance are related to the formation of a conductive path not only by electron flux, but also trap assisted tunneling phenomenona.

In fact, inducing HRS and LRS in memory cells is very convenient because that fits a real device. However, a highly selective etching has to be employed in order to remove just the electrode and not the underlying insulating layer. The etching agents must present a high selectivity with respect to the electrode material in order not to affect the underlying surface and the CFs. Moreover, the CFs observed in HRS and LRS correspond to different cells and, therefore, the properties of a single CF in both states can be assessed. This problem is solved in the second type of experiment. In this case, the tip of the CAFM is used as top electrode and a sequence of scans is performed by applying different voltages to induce the set-reset-set cycle. This is a common methodology used by many researchers. As an example, Yang et al. [36] sequentially scanned the surface of a TiO_2 stack with different voltages (Figure 3). The read voltage in that experiment was set to +1V. A SET operation in the center of a 500nm x 500nm region was simulated using a voltage of -7V (b). A subsequent read scan reveals an elevated density of leaky sites at that position and driving higher currents. Applying +7V can reset the region in the center (change the properties of the center area into HRS, d), indicating that a transition in resistance has taken place.

An important disadvantage that the CAFM brings about is the instability of the metallic tip layer, which due to high current density (determined by the tip/sample contact area) can contaminate the surface by electrochemical metallization, leading to irreversible BD events. Therefore, before starting the experiments a tip that employs a material stable enough to withstand the stresses as good as possible should be chosen. Approaches in this direction can be found for example in diamond-coated tips [39] or graphene-layered CAFM tips [40-41]. One method to overcome this drawback is to change drastically the tip design, in such a way that its bulk material is entirely made up of a conductive material like a metal or doped diamond. Regarding solid metallic tip, recent progress in fabrication technology made possible to plan and realize tip design with radii below 8nm [42]. This, in fact, is great leap ahead taking into account that handling process technology for metal bulk tips is quite more difficult than for silicon. Bulk doped-diamond tip (which are already commercially available) are also a very good option, which exhibit a superior endurance compared to metal-coated silicon and metallic bulk AFM tips. However, they are also very expensive.

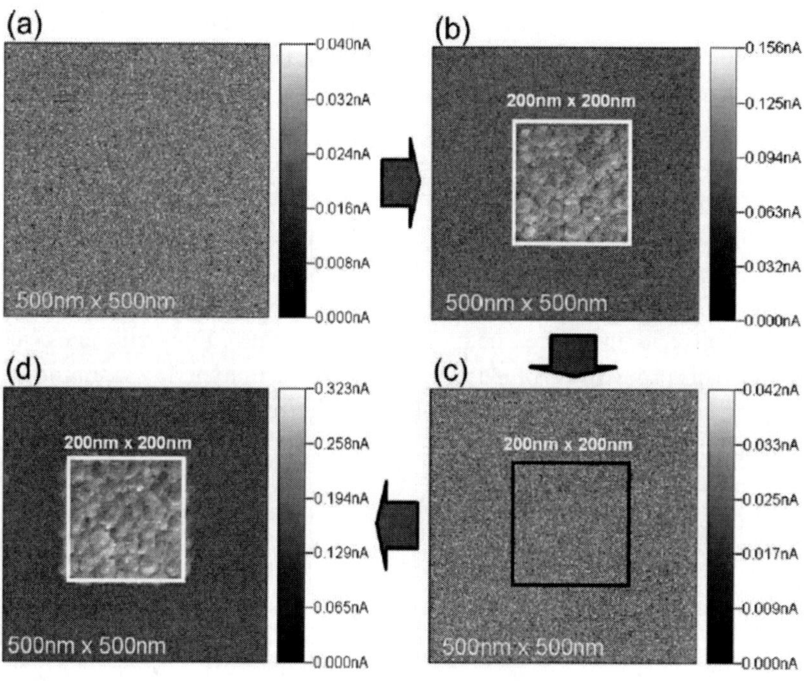

Figure 3. The current images recorded at a voltage of +1V reveal a high and low resistive state after sub sequential SET and RESET scans, respectively. The used voltage for the SET/RESET was chosen to be -7V/+7V. Reproduced with permission from [38]. Copyright [2009] AIP Publishing LLC.

Depending on the type of information we are interested to obtain, we may use the most appropriate CAFM experiment. In this sense, it is worth noting that, no one of both experiments can display the real kinetics of the RS (i. e. currents during filament rupture). The forming, set and reset processes have been typically observed by means of IV curves performed at the device level using the SPA, which can measure large currents up to miliamperes. However, investigation of RS with CAFM-based techniques involves a much more complex measurement handling procedures.

More specifically, the electrical tests oriented to observe the dynamics of RS require mainly two conditions: i) the use of a current limitation (CL) during the forming process in order to control the defect generation in the dielectric. A missing current compliance during the forming could lead to for example electrochemical metallization and Dielectric Breakdown Induced Epitaxy (DBIE) making the BD become irreversible [43-44], showing currents orders of magnitude larger than in LRS [45-46]; and ii) the application of an electrical stress to induce the reset process, which in some cases may require reverse polarities. These two requirements can be a serious challenge for a standard CAFM. The maximum voltages (currents) that are generally applied (measured) is ±10V (~nA), which is insufficient to monitor the forming, set and reset, and only allows performing read operations.

These limitations introduced by the setup lead to saturation of the CAFM electronics indicated by a horizontal line at the maximum current of the system. Linear amplifiers with a variable gain [47] do not present a solution to this problem, as they also amplify the noise in such a way that it masks effects occurring at low currents [48]. Moreover, a standard CAFM working in ambient air only allows measuring currents when injecting electrons from the substrate (in similar thin oxides) due to local anodic oxidation. In this work we solve these limitations by combining an SPA and a CAFM working in a controlled atmosphere (Figure 4a).

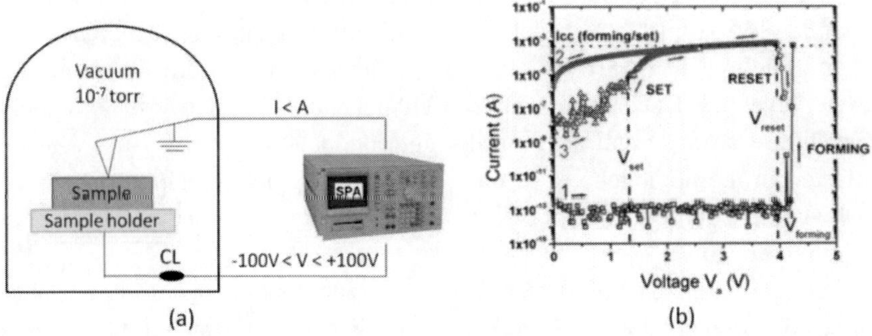

Figure 4. (a) Experimental setup for measuring bipolar filamentary resistive switching in high-k dielectrics at the nanoscale. b) IV characteristics measured in a HfO$_2$/TiN stack by biasing the AFM tip to +4V. Modified and reprinted from [27]. Published by Materials [2014]. Modified and reprinted from [52]. © IOP Publishing. Reproduced with permission. All rights reserved.

This setup was developed for the first time in prof. Nafria's group [49], and it has been extensively used to analyze pre- and post-BD electronic conduction in thin HfO2 stacks [50, 51]. One of the most recent works based on this setup, comes from [52], reporting RS in conducting nanofilaments. As figure 4b shows, both set and reset mechanisms can be clearly displayed in a current window of 12 orders of magnitude.

NANOSCALE ELECTRICAL PROPERTIES HAFNIUM OXIDES

Before analyzing RS in HfO_2 stacks, a nanoscale electrical characterization is necessary to understand the pre-BD properties of the samples. The characterization of as-grown (amorphous) hafnium-based thin insulators with CAFM has revealed that thickness fluctuations and point defects induce drastic differences on the local electrical properties [34]. In the field of HfO_2-based RRAM, moreover, it has been demonstrated that HfO2 stacks only show RS after a thermal annealing [53-54]: while the CFs formed in amorphous HfO2 layers seem to be irreversible, the thermal annealing induces important morphological changes in the insulator, leading to some regions where filament reversibility is possible. Then, the question arises on locating such special stack positions.

It is widely known that when an amorphous HfO_2 stack is subjected to a thermal annealing, pseudo-amorphous or poly-crystalline phases may be generated [55-56]. Although poly-crystallization provokes a drastic change in the morphology with respect to the amorphous phase, no drastic changes in device level pre-BD currents have been detected [57]. Nevertheless, poly-crystalline stacks present a truly different conduction characteristic, which is reflected in the homogeneity of the global current picture. These differences however are a consequence of different amounts of defects and stack thickness at grain boundaries.

As an example, Figure 5 shows the topographic and simultaneously collected current maps measured on as-grown and annealed (800°C) 5nm-HfO2/1nm-SiO$_2$/Si stacks. While the topography shows some local artifacts, the current picture is totally different: the inhomogeneities are much larger. This is produced by the polycrystallization of the sample and, despite no perfect granular structure can be observed, some grain like structures with leaky boundaries can be detected (red dashed lines in Figure 5d). Samples annealed at larger temperatures have shown perfect topography-current correlation [54]. By means of Kelvin Probe Force Microscope (KPFM) we demonstrated that GBs in the current picture correlate to higher contact potential difference (CPD) [58].

As other studies already found [59-60], stronger CPDs indicate larger concentrations of positive charges at those locations. The works of Son et al. [59] and McKenna et al. [61] give a clear indication that grain boundaries (GBs) are rich in defects, which could be filled during the KPFM mapping, leading to higher CPD signals (Figure 2d). Statistical analyses show that the deepness and current at the grain boundaries are unrelated [54], indicating that the amount of defects accumulated at the GBs is the key phenomenon that promotes the tunneling current. In addition, larger leak currents at the GBs indicate that the BD may be easier to reach at such locations (remember that the BD normally takes place at the weakest location of the sample). Therefore, annealing-related fluctuations generated at the nanoscale have a drastic effect on the BD behavior due to its local nature.

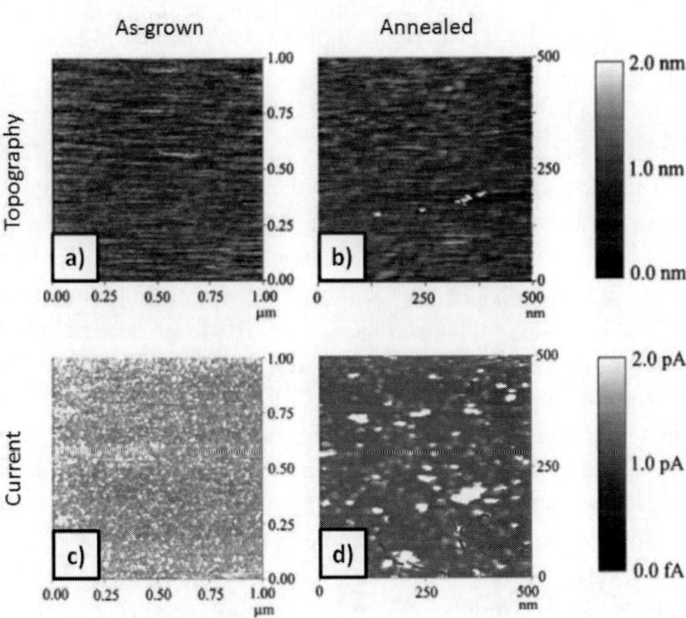

Figure 5. Topography and current maps for the sample as deposited (a and b) and annealed at 800°C (c and d), topography scale ranges from 0 to 2 nm and current from 0 to 2 fA. The images clearly show structural and electrical properties modification. Reproduced with permission from Elsevier B.V. Copyright [2004].

In summary, the CAFM and KPFM analyses revealed that polycristallization of HfO_2 leads to robust grains separated by defects-rich grain boundaries that are leakier, and also suggested that the BD event is prone to happen at the GBs. A next step to further investigate the origin of RS in HfO_2 layers should include local forming, set and reset observation from cyclic voltammograms performed at different locations of the samples, which is what we do in the next section.

PHYSICAL ORIGIN OF RESISTIVE SWITCHING IN HAFNIUM OXIDES

Once the preferred positions for the BD have been defined, sequences of IV curves of different polarities should be made to observe if the CFs can be recovered. To solve the poor voltage/current dynamic range and current limitation deficiency of commercial CAFMs, we use the setup consisting on connecting an SPA to a CAFM working in high vacuum conditions (Figure 4a). As an example, Figure 6a shows the typical IV curves measured when inducing the forming (BD) at different random locations of a polycrystalline 3 nm-thick HfO2 layer. As it can be observed, two clear different electrical behaviors can be observed. Most of the IV curves show high BD voltages above 10V (black squares). Since the area covered by the nanocrystals is much larger than the grain boundaries and since those crystals are more insulating (Figure 5d), the IV curves displayed by black squares in Figure 6a may be related to the forming inside the nanocrystals. On the contrary, a few IV curves showed forming voltages much lower (below 5V, similar to those observed at the device level [46]). Since the grain boundaries in HfO2 are leakier and since the probability that the tip was located on a grain boundary is smaller (they cover a much smaller area, around 10%), the few IV curves that showed low forming voltages may be related to the CFs formed at the GBs. Interestingly, only subsequent cyclic voltammograms performed at the GBs showed a reduction of the current (Figure 6b) similar to the reset process observed at the device level.

The data in Figure 6 indicate that the RS phenomenon is strongly linked to the nature of the grain boundaries. But, as already mentioned, the IV curves have been performed at random locations. Despite we may try to locate the tip on a GB, the intrinsic drift of the AFM together in combination with the small area covered by the GBs (they can be as narrow as 1nm), makes it very difficult to corroborate that the CAFM tip is placed just above such location. Moreover, in many samples GBs cannot be detected in topographic maps [62].

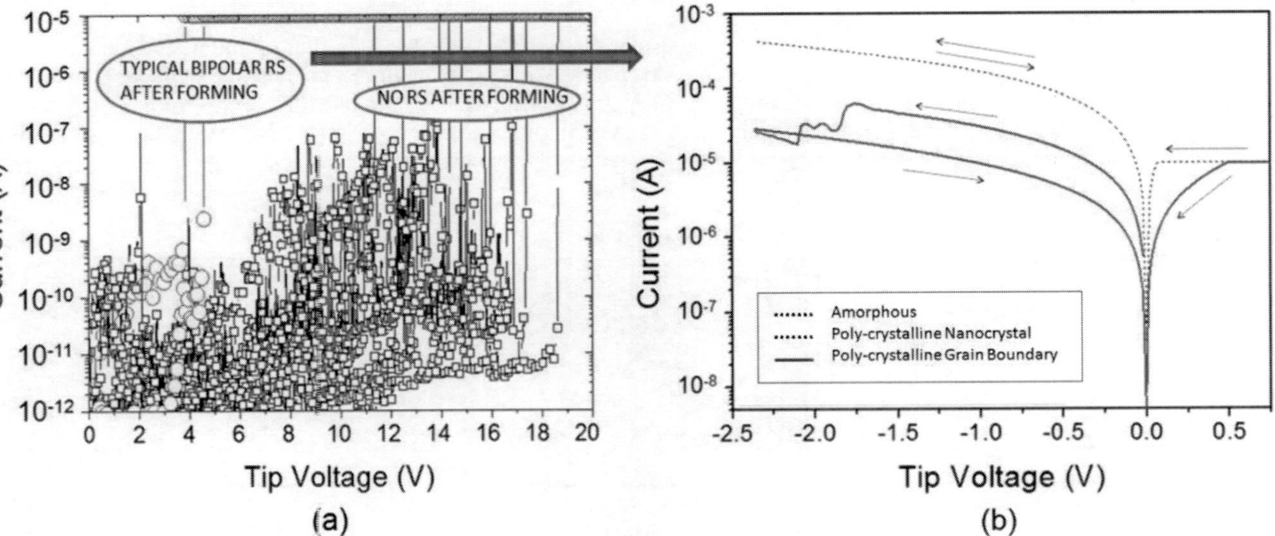

Figure 6. (a) IV curves performed at different locations of a polycrystalline 3 nm-thick HfO2 layer. Two groups of IV curves (with low and high BD voltages) can be observed. (b) typical cyclic voltammograms measured on the low and high BD voltage locations (red and dashed black lines respectively). Only low BD voltage locations show RS. Additional tests performed on amorphous samples (dashed line in b) also show irreversible BD. Modified and reprinted with permission from [46]. Copyright [2012] AIP Publishing LLC.

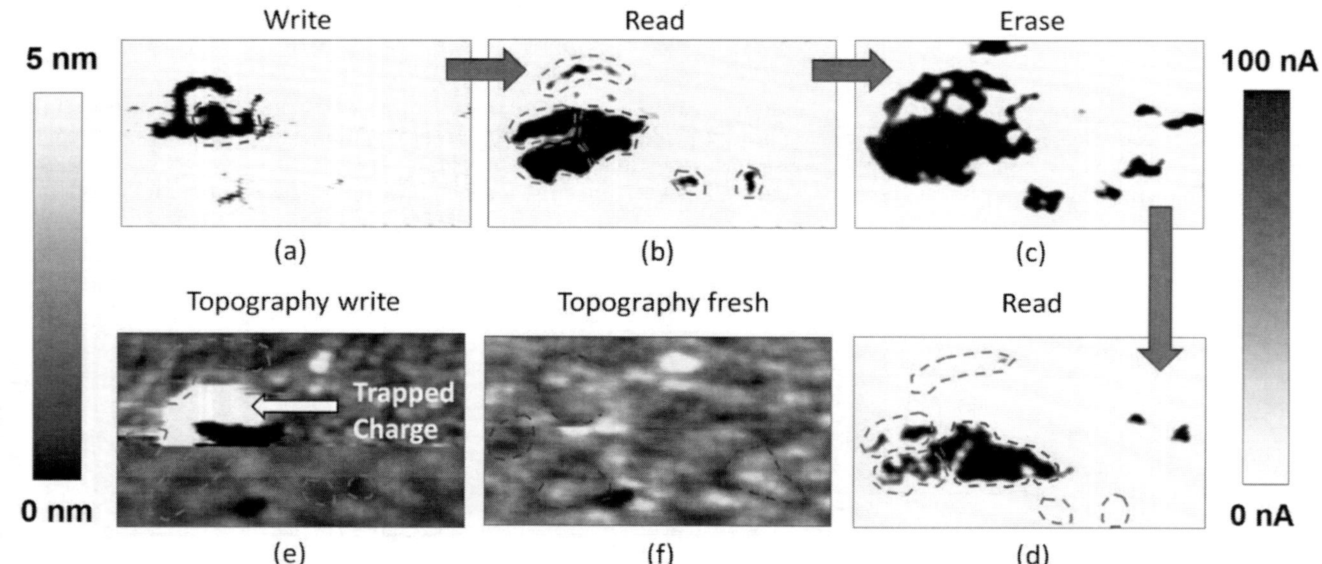

Figure 7. (a) Current and (b) topographic images collected on a polycrystalline HfO2 sample before forming. Current (c) and topographic (d) images recorded VBias=2V scan (forming). (e)–(g) exhibit sequential current images collected (after forming) by scanning at 0.3V (read), -2V (reset), and 0.3V (read), respectively. Only to induce forming, the current compliance was kept at 1μA (c). All images correspond to the same area (550 nm x 300nm). (h) and (i) show the current in multiple filaments in LRS and HRS. Modified and reprinted with permission from [45]. Copyright [2012] AIP Publishing LLC.

The analysis that may definitely prove the link between RS and GBs is a complete switching cycle (read-write-read-reset-read) induced with the tip of the CAFM in current maps. This provides morphological and current information at the same time in such a way that possible locations of leakage sites can be correlated with topographic changes. The current maps were carried out at different stages of the RS process (Figures 7a-d).

The topography picture recorded during the write process is also shown (Figure 4e) and, for comparison, the topography of a fresh sample is also displayed (Figure 4f). By applying a voltage bias of 2V, which is high enough to induce the write operation, the current spots can be observed mainly at the grains of the polycrystalline dielectric (Figure 4a), which is also indicated by a dashed line. The read process (Figure 4b) is then performed by scanning the surface of the sample with a very low voltage (0.3V) that doesn't modify the electrical properties of the stack, but only displays them. Therefore, while in Figure 4a some currents may be observed due to the tunneling current through the stack (biased at 3V), the low currents applied in Figure 4b ensure that the conductive spots observed correspond to CFs totally formed through the insulator (from bottom electrode to the stack surface). Then comes the most critical part of the experiment: destroying the conductive filaments. The erase process is performed by applying a high reverse bias of -2V (Figure 4c).

Finally another read process is induced using 0.3V. By comparing Figures 4b and 4d is can be concluded that, despite some other new conductive filaments are generated during the erase operation, the current of most of the filaments formed during the write operation (Figure 4b) is reduced after the erase operation (Figure 4d). The reversible CFs are highlighted with green dashed lines. On the contrary, the analyses also point out that some filaments are irreversible (red dashed lines), showing currents that are of the same order of magnitude.

We further analyze the origin of these CFs exploring the topographic images. Interestingly, most CFs generated during the write operation are concentrated at sample depressions (green dashed line in Figure 7e). The topographic image of the fresh sample prior to the write process further indicates that those depressions correspond to GBs (Figure 4f). It is worth noting that one of the CFs is not concentrated at the GBs, and propagates to the nanocrystal (red dashed line in Figures 7a and 7e). This formation, however, is accompanied by a large topographic artifact, which not only could originate from dielectric breakdown induced epitaxy (DBIE), but also to a large amount of charges trapped at the respective region, where the CF was formed. These positive fixed charges probably come from the metallic varnish of the tip, which under conditions of high current density and mechanical wear out, might be induced. Therefore, it can be concluded that the resistive switching in poly-crystalline HfO_2 takes place at the grain boundaries.

As demonstrated in Figure 5d and others [54], these specific locations of the sample are leakier because they contain a larger density of defects, through which current can flow by Trap Assisted Tunneling. When high voltages are applied, the dielectric breakdown in these locations can be reached at much lower voltages (Figure 6), leading to a less drastic BD event, i. e. the CFs are narrower and easier to reoxidize in subsequent cyclic voltammograms (Figure 6b).

The first models on RS in HfO_2 [63-71] suggested the possibility that the CFs are made up of oxygen vacancies. Further chemical, thermal and theoretical analyses supported this hypothesis [26, 60]. Related publications affirm the conclusion of the weaker nature of grain boundaries and underline their findings, where a concentration of positively charged oxygen vacancies is held responsible for higher leakage current [24, 72, 25].

CONCLUSION

Hafnium-based oxides are a realistic solution for many microelectronic devices, and HfO2-based thin films are now being introduced in the field of RRAM. Investigating the RS mechanism by means of Conductive Atomic Force Microscopy (CAFM) could be an excellent way to link electrical signals and morphologic features, which is necessary to study the origin of the switching mechanism. In this field, the use of an SPA connected to a CAFM working in high vacuum environment has been proved to be a very powerful tool to display the kinetics of the forming, set and reset processes at single locations of the samples.

Using this setup we have been able to conclude that RS in thin HfO_2 stacks is a local phenomenon that only takes place at the grain boundaries in polycrystalline (annealed) samples. GBs shows a leakier nature due to an unusual large density of defects that produce lower BD voltages, which generate narrower CFs that are easier to reoxidize.

REFERENCES

[1] Ribes G, Mitard J, Denais M, Bruyere S, Monsieur F, Parthasarathy C, Vincent E, Ghibaudo G. Review on high-k dielectrics reliability issues. *IEEE Transactions on Device and Material Reliablity.* 2005;5/1:5-19.

[2] Wilk G, Wallace R, Anthony J. High-k gate dielectrics: Current status and materials properties considerations. *Journal of Applied Physics.* 2001;89/10:5243-75.

[3] Mistry K, Allen C, Auth C, Beattie B, Bergstrom D, Bost M, Brazier M, Buehler M, Cappellani A, Chau R, Choi C, Ding G, Fischer K, Ghani T, Grover R, Han W, Hanken D, Hattendorf M, He J, Hicks J, Huessner R, Ingerly D, Jain P, James R, Jong L, Joshi S, Kenyon C, Kuhn K, Lee K, Liu H, Maiz J, Mcintyre B, Moon P, Neirynck J, Pae S, Parker C, Parsons D, Prasad C, Pipes L, Prince M, Ranade P, Reynolds T, Sandford J, Shifren L, Sebastian J, Seiple J, Simon D, Sivakumar S, Smith P, Thomas C, Troeger T, Vandervoorn P, Williams S, Zawadzki K. A 45nm Logic Technology with High-k+Metal Gate Transistors, Strained Silicon, 9 Cu Interconnect Layers, 193nm Dry Patterning, and 100% Pb-free Packaging. Technical Digest, *International Electron Devices Meeting.* 2007;247–250.

[4] Jotwani R, Sundaram S, Kosonocky S, Schaefer A, Andrade V, Novak A, Naffziger S. An x86-64 Core in 32 nm SOI CMOS. *IEEE Journal of Solid-State Circuits.* 2011; 46/1:162–172.

[5] Shannon R. Dielectric polarizabilities of ions in oxides and fluorides. Journal of Applied Physics. 1993;73/348.

[6] Hubbard K, Schlom D. Thermodynamic stability of binary oxides in contact with silicon. *Journal of Materials Research.* 1996;11/11:2757-76.

[7] Lee C, Hur S, Shin Y, Choi J, Park D, Kim K. Charge-trapping device structure of SiO_2/SiN/high-k dielectric Al_2O_3 for high-density flash memory. *Applied Physics Letters.* 2005; 86/152908.

[8] Ye P, Yang B, Ng K, Bude J, Wilk G, Halder S, Hwang J. GaN metal-oxide-semiconductor high-electron-mobility-transistor with atomic layer deposited Al_2O_3 as gate dielectric. *Applied Physics Letters.* 2005;86/063501.

[9] Huff H, Hou A, Lim C, Kim Y, Barnett J, Bersuker G, Brown G, Young C, Zeitzoff P, Lysaght P, Gardner M, Murto R. Integration of High-k Gate Stacks into Planar, Scaled CMOS Integrated Circuits. *Microelectronic Enginieering.* 2003;69:152-167.

[10] Extremetech. ReRAM, the memory tech that will eventually replace NAND flash, finally comes to market. [Internet] URL: http://www.extremetech.com/computing/163058-reram-the-new-memory-tech-that-will-eventually-replace-nand-flash-finally-comes-to-market.

[11] Szot K, Speier W, Bihlmayer G, Waser R. Switching the electrical resistance of individual dislocations in single-crystalline $SrTiO_3$. *Nature Materials*. 2006; 5:312-20.

[12] Beck A, Bednorz J, Gerber C. Reproducible switching effect in thin oxide films for memory applications. *Applied Physics Letters*. 2000;77:139-41.

[13] Baek I, Lee M, Seo S. Highly scalable non-volatile resistive memory using simple binary oxide driven by asymmetric unipolar voltage pulses. *Proceedings of the IEEE International Electron Devices Meeting*, San Francisco. 2004;6:587-90.

[14] Lee S, Kim W, Rhee S, Yong K. Resistance Switching Behaviors of Hafnium Oxide Films Grown by MOCVD for Nonvolatile Memory Applications. *Journal of the Electrochemical Society*. 2008;155/2:H92-96.

[15] Sawa A. Resistive switching in transition metal oxides. *Materials Today*. 2008;11:28-36.

[16] Yoshida C, Kinoshita K, Yamasaki T. Direct observation of oxygen movement during resistance switching in NiO/Pt film. *Applied Physics Letters*. 2008;93/042106:1-3.

[17] Yang J, Pickett M, Li X, Ohlberg D, Stewart D, Williams R. Memristive switching mechanism for metal/oxide/metal nanodevices. *Nature Nanotechnology*. 2008;3:429–33.

[18] Bersuker G, Gilmer D, Veksler D, Yum J, Park H, Lian S, Vandelli L, Padovani A, Larcher L, McKenna K, Shluger A, Iglesias V, Porti M, Nafria M, Taylor W, Kirsch P, Jammy R. Metal oxide RRAM switching mechanism based on conductive filament microscopic properties. *Proceedings of the IEEE International Electron Device Meeting*. 2010;19.6.1-4.

[19] Suñé J, Placencia I, Barniol N, Farrés E, Martín F, Aymerich X. On the breakdown statistics of very thin SiO_2 films. *Thin Solid Films*. 1990;185:347-62.

[20] Degraeve R, Roussel P, Groeseneken G, Maes H. A new analytic model for the description of the intrinsic oxide breakdown statistics of ultra-thin oxides. *Microelectronics Reliability*. 1996;36:1639-42.

[21] Degraeve R, Kaczer, Groeseneken G. Degradation and breakdown in thin oxide layers: Mechanisms, models, and reliability prediction. *Microelectron Reliability.* 1999; 39:1445–60.

[22] Harris R. Is ReRAM the end of NAND flash?. Storage Bits 2012.

[23] Schindler C, Thermadam S, Waser R, Kozicki M. Bipolar and unipolar resistive switching in Cu-doped SiO_2. *IEEE Transaction on Electron Devices.* 2007;54:2762-68.

[24] Bersuker G, Yum J, Vandelli L, Padovani A, Larcher L, Iglesias V, Porti M, Nafria M, McKenna K, Shluger A, Kirsch P, Jammy R. Grain boundary-driven leakage path formation in HfO_2 dielectrics. Solid-State Electronics. 2011;65–66:146–50.

[25] Long S, Perniola L, Cagli C, Buckley J, Lian X, Miranda E, Pan F, Liu M, Suñé J. Voltage and power-controlled regimes in the progressive unipolar RESET transition of HfO_2-Based RRAM. Scientific Reports. 2013;3/2929.

[26] Villena M, Jiménez-Molinos F, Roldán J, Suñé J, Long S, Lian X, Gámiz F, Liu M. An in-depth simulation study of thermal reset transitions in resistive switching memories. *Journal of Applied Physics.* 2013;114:144505.

[27] Lanza M. A review on resistive switching in thin high-k dielectrics from a nanoscale point of view. *Materials.* 2013;6.

[28] Lee M, Han S, Jeon S, Park B, Kang B, Ahn S, Kim K, Lee C, Kim C, Yoo I, Seo D, Li X, Park J, Lee J, Park Y. Electrical Manipulation of Nanofilaments in Transition-Metal Oxides for Resistance-Based Memory. *Nano Letters.* 2009;9/4:1476-81.

[29] Sun X, Sun B, Liu L, Xu N, Liu X, Han R, Kang J, Xiong G, Ma T. Resistive Switching in CeOx Films for Nonvolatile Memory Application. *IEEE Electron Device Letters.* 2009;30/4.

[30] Sung M, Kim S, Joo M, Sung J, Ryu C, Hong S, Kim H, Kim Y. Effect of the oxygen vacancy gradient in titanium dioxide on the switching direction of bipolar resistive memory. *Solid-State Electronics.* 2011;63:115–18.

[31] Shubhakar K, Pey K, Kushvaha S, O'Shea S, Raghavan N, Bosman M, Kouda M, Kakushima K, Iwai H. Grain boundary assisted degradation and breakdown study in cerium oxide gate dielectric using scanning tunneling microscopy. *Applied Physics Letters.* 2011;98:072902.

[32] Kwon D, Kim K, Jang J, Jeon J, Lee M, Kim G, Li X, Park G, Lee B, Han S, Kim M, Hwang C. Atomic structure of conducting nanofilaments

in TiO$_2$ resistive switching memory. Nature Nanotechnology. 2010;5:148–53.

[33] Uppal H, Mitrovic I, Hall S, Hamilton B, Markevich V, Peaker A. Breakdown and degradation of ultrathin Hf-based (HfO2(x)(SiO2)(1-x). *Journal Vaccuum Science and Technology B.* 2099;27:443-47.

[34] Bayerl A, Lanza M, Porti M, Nafría M, Aymerich X, Campabadal F, Benstetter G. Nanoscale and Device Level Gate Conduction Variability of High-k Dielectrics-Based Metal-Oxide-Semiconductor Structures. *IEEE Transactions on Device and Materials Reliability.* 2011;11/3.

[35] Erlbacher T, Yanev V, Rommel M, Bauer A, Frey L. Gate oxide reliability at the nanoscale evaluated by combining conductive atomic force microscopy and constant voltage stress. *Journal of Vacuum Science and Technology B.* 2011;29:01AB08.

[36] Iglesias V, Lanza M, Porti M, Nafría M, Aymerich X, Liu L, Kang J, Bersuker G, Zhang K, Shen Z. Nanoscale observations of Resistive Switching High and Low conductivity states on TiN/HfO$_2$/Pt structures. Microelectronics Reliability. 2012;52: 2110-14.

[37] Kim H, An H, Kim K, Seo Y, Nam K, Chung H, Lee E, Kim T. Large resistive-switching phenomena observed in Ag/Si3N4/Al memory cells. *Semiconductor Science and Technology.* 2010;25/6.

[38] Yang L, Kuegeler C, Szot K, Ruediger A, Waser R. The influence of copper top electrodes on the resistive switching effect in TiO$_2$ thin films studied by conductive atomic force microscopy. *Applied Physics Letters.* 2009;95:013109.

[39] Bruker AFM probes. [Internet] URL: http://www.brukerafmprobes.com/ p-3250-ddesp-fm-10.aspx

[40] Lanza M, Bayerl A, Gao T, Porti M, Nafria M, Jing G, Zhang Y, Liu Z, Duan H. Graphene-coated Atomic Force Microscope tips for reliable nanoscale electrical characterization. *Advanced Materials.* 2013;25:1440-44.

[41] Lanza M, Iglesias V, Porti M, Nafria M, Aymerich X. Polycrystallization effects on the nanoscale electrical properties of high-k dielectrics. *Nanoscale Research Letters.* 2011;6/1.

[42] Bruker AFM probes. [Internet] URL: http://www.brukerafmprobes.com/ Product.aspx?ProductID=3774

[43] Lanza M, Porti M, Nafría M, Aymerich X, Whittaker E, Hamilton B. UHV CAFM characterization of high-k dielectrics: effect of the technique resolution on the pre- and post-breakdown electrical measurements. *Microelectronics Reliability.* 2012;50:1312-15.

[44] Raghavan N, Pey K, Shubhakar K, Bosman M. Modified Percolation Model for Polycrystalline High-κ Gate Stack With Grain Boundary Defects. *IEEE Electron Device Letters*. 2011;32:78-80.

[45] Lanza M, Bersuker G, Porti M, Miranda E, Nafría M, Aymerich X. Resistive switching in hafnium dioxide layers: Local phenomenon at grain boundaries. *Applied Physics Letters*. 2012;101:193502.

[46] Lanza M, Zhang K, Porti M, Nafria M, Shen Z, Liu L, Kang J, Gilmer D, Bersuker G. Grain boundaries as preferential sites for resistive switching in the HfO$_2$ resistive random access memory structures. *Applied Physics Letters*. 2012;100:123508.

[47] Femto Amplifiers. [Internet] URL: http://www.femto.de/en/products/current-amplifiers/variable-gain-up-to-200-mhz-dhpca.html

[48] Lanza M, Porti M, Nafría M, Aymerich X, Benstetter G, Lodermeier E, Ranzinger H, Jaschke S, Wilde L, Michalowski P. Conductivity and charge trapping after electrical stress in amorphous and polycristalline Al$_2$O$_3$ based devices studied with AFM related techniques. *IEEE Transactions on Nanotechnology*. 2011;10:344-51.

[49] Blasco X, Pétry J, Nafría M, Aymerich X, Richard O, Vandervorst W. C-AFM characterization of the dependence of HfAlOx electrical behavior on post-deposition annealing temperature. *Microelectronic Engineering*. 2004;72/1–4:191–6.

[50] Aguilera L, Lanza M, Bayerl A, Porti M, Nafria M, Aymerich X. Development of a conductive atomic force microscope with a logarithmic current-to-voltage converter for the study of metal oxide semiconductor gate dielectrics reliability. *Journal of Vacuum Science and Technology B*. 2009;27/1. doi: http://dx.doi.org/10.1116/1.3021049.

[51] Uppal H, Markevich V, Volkos S, Dimoulas A, Hamilton B, Peaker A. Post-stress/breakdown leakage mechanism in ultrathin high-κ (HfO$_2$)x(SiO$_2$)1-x/SiO$_2$ gate stacks: A nanoscale conductive-Atomic Force Microscopy C-AFM. *MRS Proceedings*. 2008;1108. doi:http://dx.doi.org/10.1557/PROC-1108-A10-06.

[52] Calka P, Martinez E, Delaye V, Lafond D, Audoit G, Mariolle D, Chevalier N, Grampeix H, Cagli C, Jousseaume V, Guedj C. Chemical and structural properties of conducting nanofilaments in TiN/HfO$_2$-based resistive switching Structures. *Nanotechnology*. 2013;24:085706.

[53] Lanza M, Bersuker G, Porti M, Miranda E, Nafría M, Aymerich X. Resistive switching in hafnium dioxide layers: Local phenomenon at grain boundaries. *Applied Physics Letters*. 2012;101:193502.

[54] Iglesias V, Porti M, Nafria M, Aymerich X, Dudek P, Schroeder T, Bersuker G. Correlation between the nanoscale electrical and morphological properties of crystallized hafnium oxide-based metal oxide semiconductor structures. *Applied Physics Letters.* 2010;97:262906.

[55] Jeong S, Lee H, Kim K, You M, Roh Y, Noguchi T, Xianyu W, Jung J. Effects of annealing temperature on the characteristics of ALD-deposited HfO_2 in MIM capacitors. *Thin Solid Films.* 2006;515:526–30.

[56] Gusev P, Cabral J, Copel M, Emic C, Gribelyuk M. Ultrathin HfO2 films grown on silicon by atomic layer deposition for advanced gate dielectrics applications. *Microelectronic Engineering.* 2003;69:145–51.

[57] Kim H, McIntyre P. Effects of crystallization on the electrical properties of ultrathin HfO2 dielectrics grown by atomic layer deposition. *Applied Physics Letters.* 2003;82:106–8.

[58] Iglesias V, Lanza M, Porti M, Nafría M, Aymerich X, Liu L, Kang J, Bersuker G, Zhang K, Shen Z. Nanoscale observations of Resistive Switching High and Low conductivity states on TiN/HfO_2/Pt structures. *Microelectronics Reliability.* 2012;52:2110-4.

[59] Son J, Shin Y. Direct observation of conducting filaments on resistive switching of NiO thin films. *Applied Physics Letters.* 2008;92:222106.

[60] Bersuker G, Gilmer D, Veksler D, Kirsch D, Vandelli L, Padovani A, Larcher L, McKenna K, Shluger A, Iglesias V, Porti M, Nafría M. Metal oxide resistive memory switching mechanism based on conductive filament properties. *Journal of Applied Physics.* 2011;110:124518.

[61] McKenna K, Shluger A. The interaction of oxygen vacancies with grain boundaries in monoclinic HfO_2. *Applied Physics Letters.* 2009; 95/22:222111.

[62] Petry J, Vandervorst W, Blasco X. Effect of N2 anneal on thin HfO_2 layers studied by conductive atomic force microscopy. *Microelectronic Engineering.* 2004;72:174–9.

[63] Xu N, Gao B, Liu L, Sun B, Liu X, Han R, Kang J, Yu B. A Unified Physical Model of Switching Behavior in Oxide-Based RRAM. *Proceedings of the Symposium on VLSI Technology.* 2008:100–1.

[64] Gao B, Sun B, Zhang H, Liu L, Liu X, Han R, Kang J, Yu B. Unified physical model of bipolar oxide-based resistive switching memory. *IEEE Electron Device Letters.* 2009;30:1326-8.

[65] Chen B, Gao B, Sheng S, Liu L, Liu X, Chen Y, Wang Y, Han R, Yu B, Kang J. A novel operation scheme for oxide-based resistive-switching

memory devices to achieve controlled switching behaviors. *IEEE Electron Device Letters*. 2011;32:282–4.

[66] Zhang H, Liu L, Gao B, Qiu Y, Liu X, Lu J, Han R, Kang, J, Yu B. Gd-doping effect on performance of HfO2 based resistive switching memory devices using implantation approach. *Applied Physics Letters*. 2011;98:042105:1–3.

[67] Gao B, Kang J, Liu L, Liu X, Yu B. A physical model for bipolar oxide-based resistive switching memory based on ion-transport-recombination effect. *Applied Physics Letters*. 2011;98:232108:1–3.

[68] Gao B, Kang J, Chen Y, Zhang F, Chen B, Huang P, Liu L, Liu X, Wang Y, Tran X, Wang Z, Yu H, Chin A. Oxide-based RRAM: Unified microscopic principle for both unipolar and bipolar switching. In *Proceedings of the IEEE International Electron Devices Meeting. 2011.* doi: http://dx.doi.org/10.1109/IEDM.2011.6131573.

[69] Lu Y, Gao B, Fu Y, Chen B, Liu L, Liu X, Kang J. A simplified model for resistive switching of oxide based resistive random access memory devices. *IEEE Electron Device Letters*. 2012;33:306–8.

[70] Gao B, Bing C, Zhang F, Liu L, Liu X, Kang J, Yu H, Yu B. A novel defect-engineering-based implementation for high performance multilevel data storage in resistive switching memory. *IEEE Transactions on Electron Devices*. 2013;60:1379-83.

[71] Bing C, Jin F, Gao B, Ye X, Li F, Xiao Y, Zheng F, Yu Y, Xin P, Guo Q, Dim L. Endurance degradation in metal oxide-based resistive memory induced by oxygen ion loss effect. *IEEE Electron Device Letters*. 2013;34/10:1292–4.

[72] Celano U, Goux L, Belmonte A, Schulze A, Opsomer K, Detavernier C, Richard O, Bender H, Jurczak M, Vandervorst W. Conductive-AFM tomography for 3D filament observation in resistive switching devices. *IEEE International Electron Device Meeting*. 2013. doi: http://dx.doi.org/10.1109/IEDM.2013.6724679.

In: Hafnium
Editor: HongYu Yu

ISBN: 978-1-63463-164-8
© 2015 Nova Science Publishers, Inc.

Chapter 6

ULTRATHIN HAFNIUM-BASED HIGH-K DIELECTRICS FOR HIGH-K-LAST/ GATE-LAST CMOS INTEGRATION SCHEME

Shu Xiang Zhang[1], Jiang Yan[1]* and HongYu Yu[2]

[1]Institute of Microelectronics of Chinese Academy of Sciences,
ChaoYang District, Beijing, China
[2]South University of Science and Technology of China,
Nanshan District, Shenzhen, Guangdong, China

ABSTRACT

To fulfill the scaling scenario of advanced CMOS devices as projected by the International Technology Roadmap for Semiconductors, high quality ultrathin gate dielectric has been one of the major challenges to be conquered. Currently, the mainstream in the industry employs the atomic layer deposition (ALD) method to deposit thin dielectric films, with a prime emphasis on Hf-based high-K dielectrics. In this chapter, we review our recent progress on research of ultrathin Hf-based high-K/metal gate stack technology. Especially, the HfO2 thin films deposited by ALD using TEMAH and H_2O as precursors under 300°C deposition temperature for both high-K-last and metal gate-last integration manner are investigated by the physical characterization (X-ray Photoelectron Spectroscopy, High Resolution TEM combined with Electron Energy

---

* Corresponding Author Email: yanjiang@ime.ac.cn

Loss Spectroscopy), and the electrical characterization (capacitance voltage – CV & current voltage – IV measurements) techniques. The corresponding device characteristics are discussed to illustrate the complex issues for high-K dielectric integration into current CMOS technology.

Keywords: HfO_2, Gate-last, Equivalent oxide thickness (EOT)

INTRODUCTION

Thermally grown silicon dioxide (SiO_2) has been consistently used in CMOS technologies for the past few decades. SiO_2 with physical thickness of 1.2 nm has been implemented in the 90-nm logic technology node and CMOS transistors with 0.8-nm (physical thickness) SiO_2 have also been demonstrated [1]. However, with the continuous decrease of gate oxide thickness, gate leakage current has become a serious concern to limit its application. Thus, high-K materials, such as HfO_2, ZrO_2, Al_2O_3, Ta_2O_5, and TiO_2 have been intensively studied to replace SiO_2. These high-K materials can maintain the same gate capacitance as SiO_2 but with a much thicker physical thickness, thus leading to the reduction of the gate leakage current.

Among various high-K materials mentioned above, HfO_2 has emerged as the leading candidate to replace SiO_2 due to its relatively high dielectric constant (~20), reasonable band alignment with Si substrate, and more importantly the compatibility with current CMOS processes [2]. To minimize the related defects in the transition region from the Si substrate to the gate dielectric, a thin interfacial layer (IL) is usually grown prior to the deposition of HfO_2 thin films [3]. On the other hand, it is noted that the IL growth shall increase the final equivalent oxide thickness (EOT) of the gate stack, i.e., reduce the gate capacitance. Similar to other high-K materials, HfO_2 can also be deposited by several means, such as atomic layer deposition (ALD), metal organic chemical vapor deposition (MOCVD), and physical vapor deposition (PVD). Among them, ALD has been extensively employed for HfO_2 thin films deposition. During ALD, materials are deposited layer by layer in a self-limiting manner, allowing for inherent atomic scale control. The ALD process can find its advantages in accurate thickness control and excellent step coverage because the growth rate only depends on the number of growth cycles and the lattice parameters of materials.

In this chapter, our recent research progress on HfO_2 thin films deposited by ALD for integration in the high-K-last/gate-last CMOS process is reviewed. To achieve sub-1 nm EOT, interfacial oxide formation suppression is achieved via "oxygen-scavenging effect" using Ti metal capping on HfO_2 dielectric with an appropriate annealing process. Meanwhile TiN is employed as a diffusion barrier layer underneath Ti metal to protect high-K dielectric from possible degradation induced by Ti penetration into HfO_2, which is important to maintain the HfO_2 integrity. Further, a multi deposition multi annealing (MDMA) technique is introduced to reduce gate leakage current via healing of oxygen vacancies in HfO_2 films and IL, and the results are compared to the reference gate stack treated by conventional one-time deposition and annealing (D&A).

EXPERIMENTAL

The MOS capacitor fabrication process flow is illustrated in Figure 1. After ~300 nm thermal field oxide growth on the n-type (100) Si substrate ~$10^{15}cm^{-2}$) and active area definition, ~1 nm chemical oxide IL is grown by an O_3 de-ionized (DI) water cleaning to improve the interface quality between high-K and substrate.

Then the wafer is placed in the ALD chamber. The precursors used for the HfO_2 thin films deposition are TEMAH [tetrakis(ethylmthylamino)-hafnium] and H_2O. The HfO_2 thin films are deposited by alternating pulses of TEMAH and H_2O. A deposition cycle is thus defined as a pulse of TEMAH, an argon purge process (to remove the unreacted precursor), an H_2O pulse, and another argon purge process.

The deposition temperature is kept as 300°C and the HfO_2 thin film thickness is controlled by the number of cycles. After ALD deposition, the films then receive post-deposition-annealing (PDA), i.e., are annealed at 450°C in a N_2/O_2 [~10/1] ambient for 15s. Finally, TiN based gate electrode are deposited and patterned.

After metallization and alloy annealing process, the devices are then tested. The analysis of interfacial region of HfO_2/IL is done using XPS (X-ray photoelectron spectrometry), TEM (transmission electron microscopy) and EELS (electron energy loss spectroscopy).

The electrical characteristics are evaluated using a Keithly 4200-SCS semiconductor parameter analyzer. The capacitance–voltage curves are measured at 1MHz to determine the EOT and flat-band voltage using a quantum-mechanical CV simulator (CVC program).

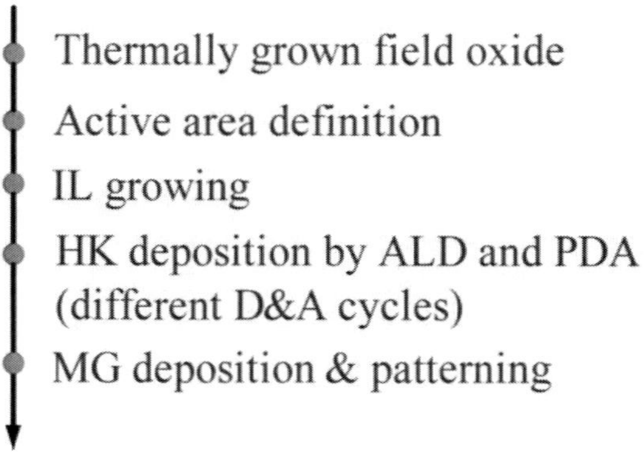

Thermally grown field oxide

Active area definition

IL growing

HK deposition by ALD and PDA
(different D&A cycles)

MG deposition & patterning

Figure 1. MOS capacitor fabrication process flow.

PHYSICAL CHARACTERIZATION OF HfO_2/SiO_2/Si STACK

Figure 2 shows the XPS spectra of Si $2p$ and Hf $4f$ for the as-deposited HfO_2/SiO_2/Si structure with different thicknesses of HfO_2. It should be noted that these samples do not receive PDA treatment in order to prevent the possible compositional intermixing of the whole HfO_2/SiO_2/Si stack. The Hf $4f$ peak is observed in range of 12 ~ 15 eV, which represents the formation of Hf-O bonds [4]. It can be seen from Figure 2 that all the spectra of Si $2p$ can be well-deconvoluted into two peaks. One peak is located at 99.3 eV originating from Si substrate (Si $2p$Si_sub), and the other is located at higher binding energy of ~103 eV originating from SiO_2 (Si $2p$SiO$_2$) [5]. The spectra of Hf $4f$ can be well-fitted by one peak, indicating only one chemical state originating from HfO_2 (Hf $4f$HfO$_2$), and there is no intermixing between HfO_2 and SiO_2.

Figure 3 shows the cross-sectional HRTEM image of the HfO_2/SiO_2/Si structure after 450°C PDA. The thickness of the HfO_2 film is ~ 2.2 nm ~1 nm interfacial layer between HfO_2 and the silicon substrate can be clearly seen. From XPS (data not shown), it is also found that interfacial mixing occurs at the interface between HK and IL, which is due to the PDA thermal budget. Such a mixing thus leads to the formation of Hf-silicate (Si–O–Hf). With more Hf-O bonds, the chemical oxide IL improves the HK/IL interface and thus enhances the quality of the subsequent deposited high-K layer. It has been reported that the formation of such Hf-silicate should provide a larger effective K value and thus a smaller EOT can be achieved as compared to the pure SiO_x IL.

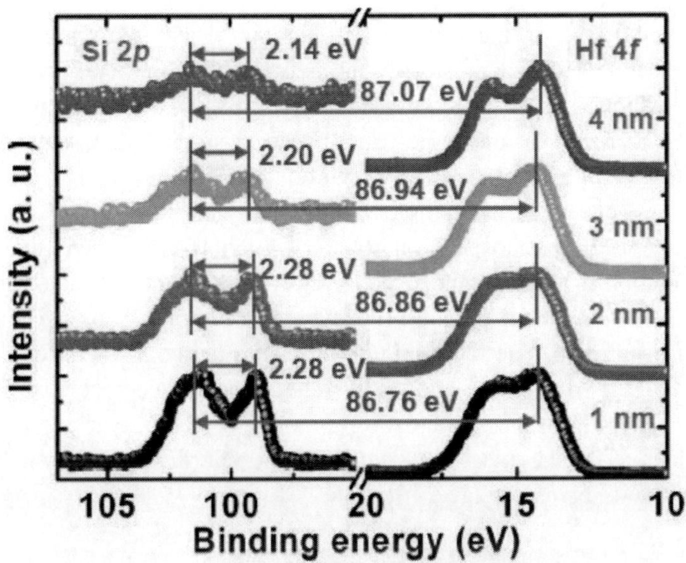

Figure 2. XPS results of Si 2p and Hf 4f of $HfO_2/SiO_2/Si$ structure with different thicknesses of HfO_2.

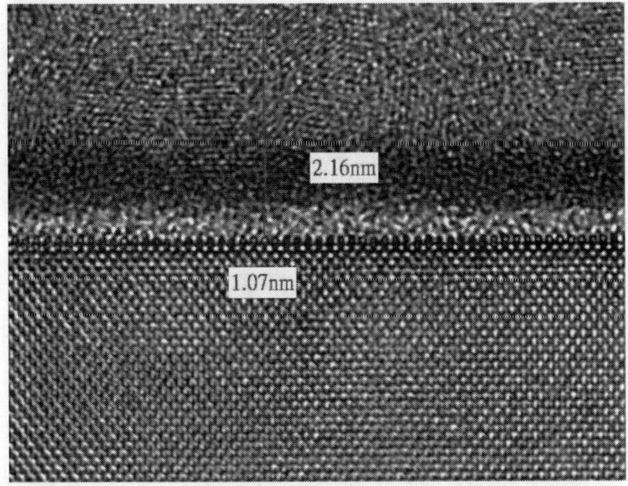

Figure 3. HRTEM images of the $HfO_2/SiO_2/Si$ structure.

REMOTE SCAVENGING TECHNOLOGY USING A TI SCAVENGING LAYER AND A TIN BARRIER LAYER ON HFO$_2$ DIELECTRIC

Figure 4 displays the schematic diagram of oxygen-scavenging gate structure. The Ti/TiN layers are inserted between the top metal gate electrode and HfO$_2$, which realizes the oxygen-scavenging effect and also suppresses the possible Ti diffusion at the same time. Figure 5 shows cross-sectional TEM images of the gate stacks with and without Ti/TiN layers respectively. It should be noted that all the samples are treated with rapid thermal annealing (RTA) process in N$_2$ atmosphere at 600°C to promote the scavenging reaction. The scavenging approach indeed achieves a thinner interfacial oxide layer. This result suggests that the thin Ti layer absorbs excess oxygen and thus inhibits IL re-growth. The scavenging effect is further confirmed by Si $2p$ XPS analysis (Figure 6), where the SiO$_2$ related peak becomes smaller after oxygen-scavenging process, indicating that the Ti metal layer competes with the Si substrate for the excess oxygen absorption during anneal, which results in thinner low-k interfacial layer [6].

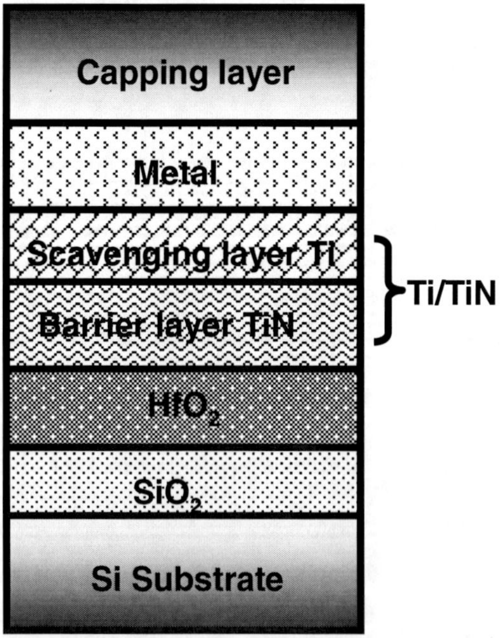

Figure 4. Schematic diagram of the oxygen-scavenging structure.

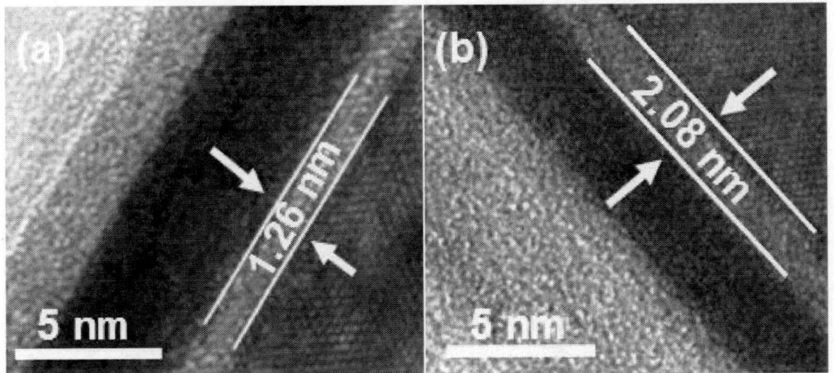

Figure 5. HRTEM of samples with (a) and without (b) Ti/TiN, after 600°C RTA.

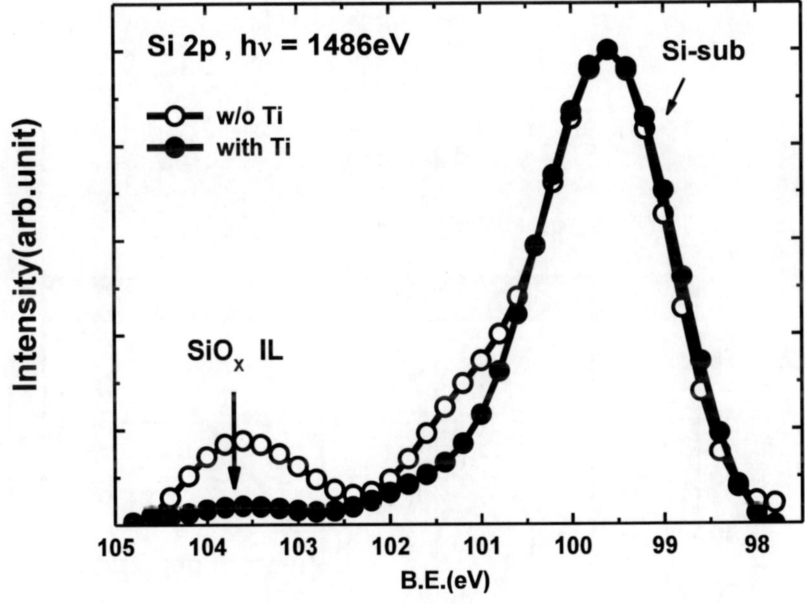

Figure 6. XPS spectra of samples with and without the Ti/TiN bi-layer after 600°C RTA. Open circle and closed circle represent samples without and with Ti/TiN respectively.

Furthermore, the effect of the barrier layer TiN on preventing metal Ti diffusing into high-K material is confirmed by an EELS depth profile analysis (Figure 7 and Figure 8). As compared with gate structure without the TiN barrier layer (Figure 8), penetration of Ti into the HfO_2 dielectric upon annealing in the gate structure with the TiN layer (Figure 7) has been effectively suppressed, which is highly beneficial to maintain the HfO_2 integrity.

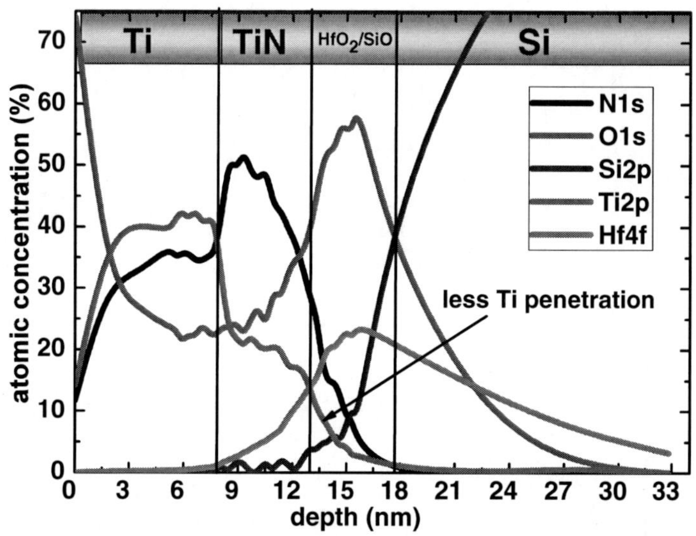

Figure 7. Depth profile of sample with TiN barrier layer.

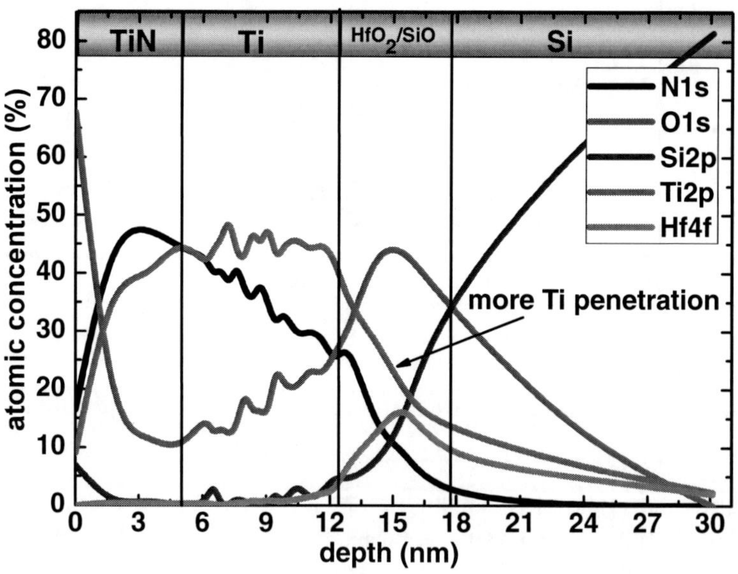

Figure 8. Depth profile of sample without TiN barrier layer.

HIGH-K PERFORMANCE IMPROVEMENT THROUGH MULTI-DPOSITION-MULTI-ANNEALING (MDMA) TECHNIQUE

In a high-K last process, due to the limitation on the thermal budget (normally <600 $^{\circ}$C to avoid the degradation of NiSi source/drain), the one-time annealing can only modify the very top surface materials in a relatively thick layer, and as a result the defects in deep bulk HfO_2 and IL can't be cured effectively. In the case of MDMA, the annealing step is programmed after cycles of ALD are given; so that oxygen vacancies can thus be effectively healed in deep bulk high-K films and also IL. However, MDMA can lead to EOT increase since more oxygen has a chance to react with the interfacial layer. Thus in our case, a Ti scavenging layer is kept in the gate stack. In the following text, we name the samples with Ti scavenging layer as process A, and the samples without Ti scavenging layer as process B. For each process, four samples with a different cycle number of deposition and annealing (D&A) are prepared as shown in Table 1: 15s×1, 15s×2, 15s×4, 15s×7, which refers to 1, 2, 4, 7 D&A cycles. The total ALD deposition cycles are kept as constant among all samples to keep the physical thickness of HfO_2 unchanged.

Table 1. The details of high-k deposition and annealing conditions for all samples

Sample	D&A Times	ALD HfO_2 cycles/time	Annealing after each time	Total HK cycles	Total annealing time
PDA 15s×1	1	28	RTA 450°C 15S N_2/O_2	28	15s
PDA 15s×2	2	14	RTA 450°C 15S N_2/O_2	28	30s
PDA 15s×4	4	7	RTA 450°C 15S N_2/O_2	28	60s
PDA 15s×7	7	4	RTA 450°C 15S N_2/O_2	28	105s

The HfO_2 film treated with only one-time PDA is initially studied, and the deformed CV curve, owing to the high leakage current, is revealed in Figure 9(a). We can conclude that the PDA 15s×1 sample exhibits a high number of defects [8]. This is because one-time D&A treatment can only modify the very top surface quality, which means the bulk HfO_2 may still maintain the low quality inherent from the as-deposited film.

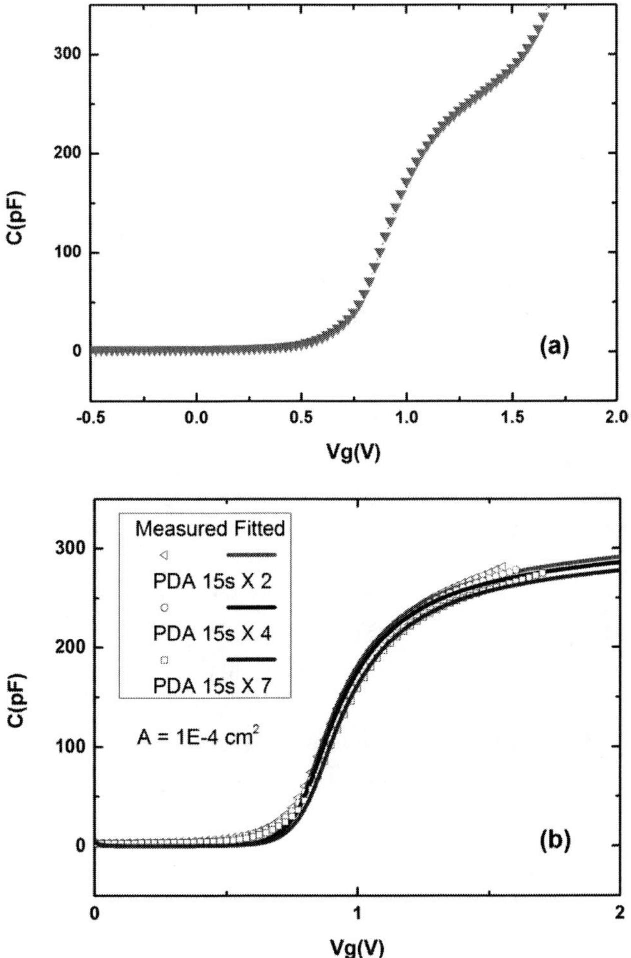

Figure 9. (a) Measured CV for devices with only one-time PDA after HK deposition, devices shows high leaky behavior. (b) Measured (symbols) and fitted (lines) CV curves for devices with MDMA [7].

In Figure 9(b), MDMA devices display a well-behaved CV curve, which perfectly matches the CVC simulation with quantum mechanical correction, indicating good quality of the HK film.

The key mechanism behind MDMA improvement is illustrated in Figure 10. The conventional high-K fabrication annealing shall only modify material quality at the very top surface, such that the defects in deep bulk HfO_2 and IL might still exist. In the case for MDMA process, the annealing step is programmed after given cycles of ALD, allowing oxygen vacancy related defects to be effectively healed in bulk high-K films as well as in the interfacial layer.

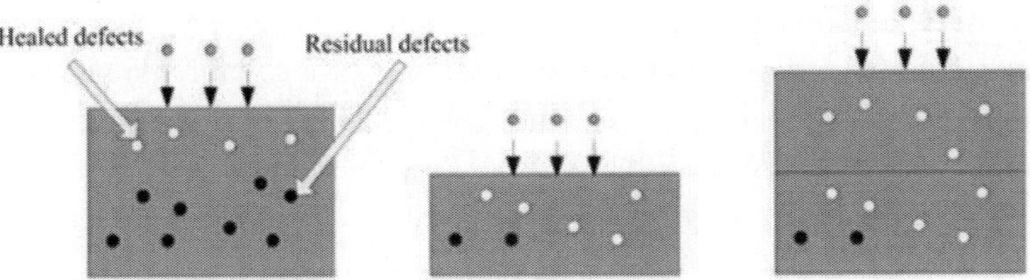

Figure 10. Schematic showing how MDMA can effectively improve HK film quality [7].

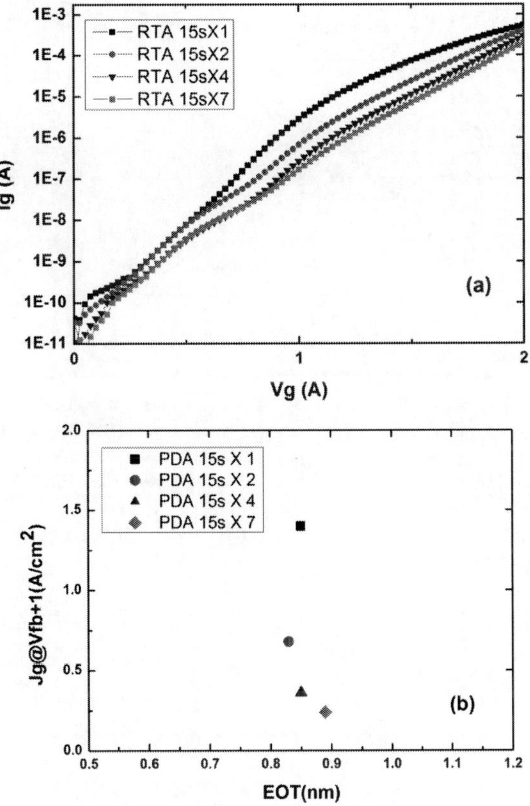

Figure 11. Process A: (a) Measured gate leakage current at room temperature. (b) Current density versus EOT at room temperature [7].

As seen in Figure 11(a) & Figure 12(a), undesirable EOT regrowth is a concern in process B as D&A times increase but not a concern in process A, mainly due to the Ti scavenging layer. Gate leakage current (Jg) measurement of process A and B is compared in Figure 11(b) and Figure 12(b) respectively. It is observed that Jg decreases as the cycle number of D&A increases. However, the reduction level tends to saturate after a certain number of D&A. We thus conclude that although MDMA reduces leakage current through healing of oxygen vacancies, the D&A cycle number should be controlled. As seen from PDA 15s×7 samples of process A, Jg reduction is insignificant as compared with the PDA 15s×4 sample, but already with a penalty on EOT increase, even using the Ti scavenging layer.

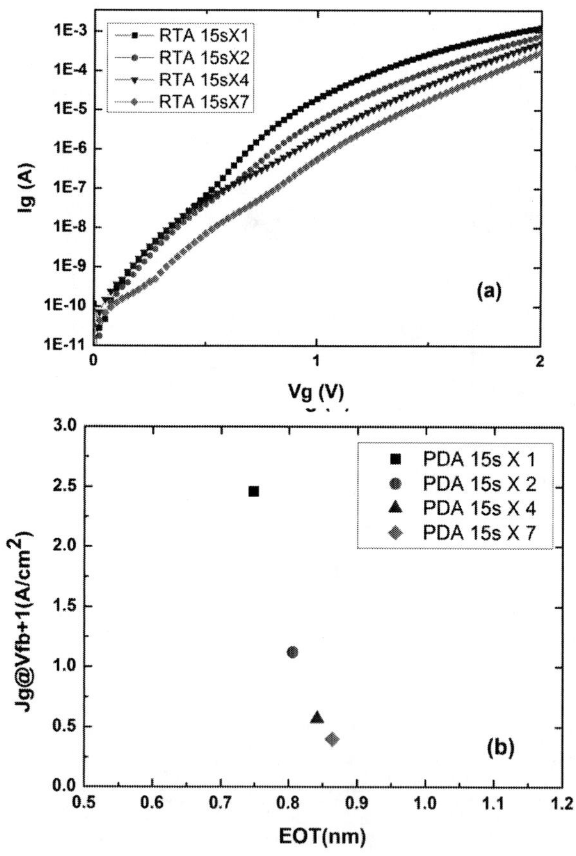

Figure 12. Process B: (a) Measured gate leakage current at room temperature. (b) Current density versus EOT at room temperature [7].

Figure 13 shows the EDX spectrum of the HK film. Oxygen content in the sample with one-time D&A is the lowest, which is possibly responsible for insufficient passivation of oxygen vacancies. With the increase of the D&A cycle numbers, more oxygen vacancies can be healed owing to more oxygen incorporation. Several research groups have shown that the K value of IL is significantly greater than that of bulk stoichiometric SiO_2, which is also verified by our data.

As shown in Figure 14, the EOT (~0.9nm) of MDMA devices in process A is even less than the IL physical thickness. As discussed previously, one explanation is that the high-K film intermixing with the underlying SiO_2 layer forms Hf silicate. Oxygen deficiency in IL is also reported to account for the K value increase. However, as seen in our case, oxygen content in IL is increased with D&A cycle number, but the K value of IL doesn't change noticeably as the extracted EOT remains almost constant for all samples in process A. This suggests that the K value boost for IL may not result from oxygen deficiencies. Anyway, with MDMA, it is expected that both oxygen vacancies in the deep bulk HK layer and IL can be cured, which would contribute to the improvement of HK performance as oxygen vacancy related defects play an important role for HK gate stack degradation.

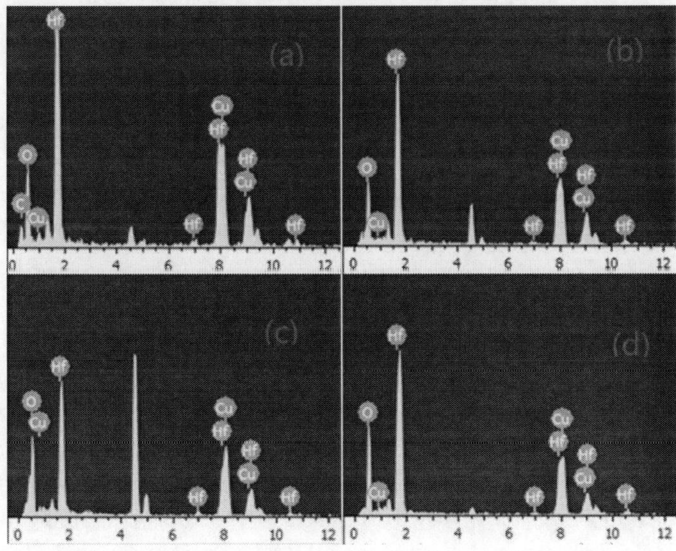

Figure 13. EDX spectrum of IL of the samples with (a) PDA 15s×1, oxygen contents 11.6%. (b) PDA 15s×2, oxygen content 14.10%. (c) PDA 15s×4, oxygen contents 20.72% (d) PDA 15s×7, oxygen content 23.27% [7].

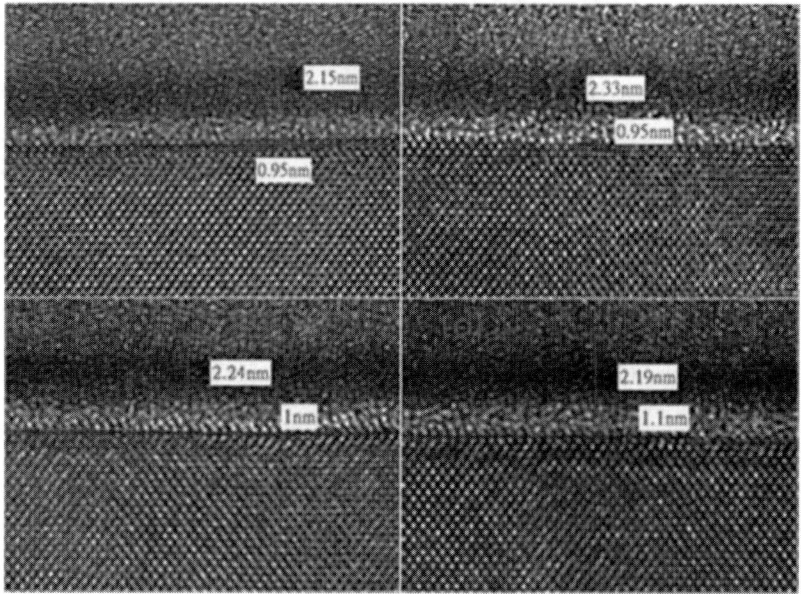

Figure 14. TEM pictures confirm that all samples in process A annealed under different conditions have similar EOT, with ~1 nm IL and ~2.2 nm HK. (a) PDA 15s×1. (b) PDA 15s×2. (c) PDA 15s×4. (d) PDA 15s×7 [7].

From the TEM micrographs (Figure 14), all samples in process A are with ~1nm IL and ~2.2 nm HfO_2 HK, regardless of annealing conditions. The IL thickness stability in process A can be attributed to the oxygen scavenging effect of the Ti layer, which possesses high oxygen solubility as well as negative free energy of oxide formation. These properties can be further used to suppress the IL regrowth and meanwhile, improves gate stack integrity.

CONCLUSION

Our recent research progress on the ultrathin HfO_2 thin films deposited by ALD integrate-able in a high-K-last/gate-last manner, is presented. A promising approach combining the oxygen-scavenging layer with a barrier layer is introduced to IL regrowth during PDA while maintaining the HfO_2 integrity at the same time. Further, the MDMA process is verified to improve HfO_2 high-K quality, which might stem from the healing of oxygen vacancies in both the deep bulk HK layer and the interfacial layer.

REFERENCES

[1]   Thompson S., Anand N., Armstrong M., Auth C., Arcot B., Alav M. & Bohr M. (2002, December). A 90 nm logic technology featuring 50 nm strained silicon channel transistors, 7 layers of Cu interconnects, low k ILD, and 1um2 SRAM cell. *Electron Devices Meeting*, 2002. IEDM'02. International (pp. 61-64). IEEE.

[2]   Choi R., Onishi K., Kang C. S., Gopalan S., Nieh R., Kim Y. H. & Lee J. C. (2002, December). Fabrication of high quality ultra-thin HfO2 gate dielectric MOSFETs using deuterium anneal. *Electron Devices Meeting*, 2002. IEDM'02. International (pp. 613-616). IEEE.

[3]   Chen Y. T., Fu S. I., Chiang W. T., Lin C. T., Tsai S. H., Wang S. W. & Chang S. J. (2012). Chemical Oxide Interfacial Layer for the High-k-Last/Gate-Last Integration Scheme. Electron Device Letters, *IEEE*, 33(7), 946-948.

[4]   Xiang J., Wang X., Li T., Zhao C., Wang W., Li J. & Ye T. (2013). Band Lineup Issues Related with High-k/SiO2/Si Stack. *ECS Transactions*, 50(4), 293-298.

[5]   Wang X., Han K., Wang W., Xiang J., Yang H., Zhang J. & Ye, T. (2012). Band alignment of HfO_2 on SiO_2/Si structure. *Applied Physics Letters,* 100(12), 122907.

[6]   Ma X., Wang X., Han K., Wang W., Yang H., Zhao C. & Ye T. (2013). Remote scavenging technology using Ti/TiN capping layer interposed in a metal/high-k gate stack. *ECS Transactions*, 50(4), 285-290.

[7]   Zhang S., Yang H., Tang B., Tang Z., Xu Y., Xu J. & Yan J. (2014). Combining multi deposition multi annealing technique with scavenging (Ti) to improve high-k/metal gate stack performance or a gate-last process. *Journal of Semiconductors*, 35(10).

[8]   Wu L., Yew K. S., Ang D. S., Liu W. J., Le T. T., Duan T. L. & Yu, H. Y. (2010, December). A novel multi deposition multi room-temperature annealing technique via Ultraviolet-Ozone to improve high-K/metal (HfZrO/TiN) gate stack integrity for a gate-last process. *Electron Devices Meeting* (IEDM), 2010 *IEEE International* (pp. 11-6). IEEE.

INDEX

N

O

S

T

U

V

W